SOIL-WATER ENGINEERING
FIELD AND LABORATORY MANUAL

Other Related Books

Soil Survey Laboratory Methods Manual	*USDA*
Standard Methods for Analysis of Soil Plant and Water	*I.C. Gupta et al.*
A Text Book of Environmental Science	*Dr. Vidya Thakur*
New Vistas of Organic Farming	*Mukund Joshi*
Soil-Water Engineering Field & Laboratory Manual	*Thomas J. Trout*
Basic Ground-Water Hydrology	*U.S.D.I.*
Indira Agriculture Competition Explorer	*R.L. Arya et al.*
A Text Book on Farming Systems	*U.K. Behera*
The Soil – Plant Systems in Relation to Inorganic Nutrition	*Maurice Fried*
Climate Change and its Ecological Implications for the Western Himalaya	*V.L. Chopra*
Plant Analysis: Comprehensive Methods and Protocols	*B.K. Garg*
Salinity Tolerance in Plants: Methods, Mechanisms & Management	*B.K. Garg*
Soil Survey Manual - Revised and Enlarged Ed.	*USDA*
Allelopathy in Soil Sickness	*S.S. Narwal*
Resource Conserving Techniques in Crop Production	*A.R. Sharma*
Soil and Plant Analysis Laboratory Manual	*John Ryan et al.*
Principles in the Quantitative Analysis of Waters, Fertilizers, Plants Soils	*U.S. Sree Ramulu*
Use of Saline Water in Agriculture	*I.C. Gupta*
Soil and Water Conservation in Semi-arid area	*FAO*
Soil Salinity Assessment Methods & Interpretation of Electrical Conductivity Measurements	*FAO*
Use of Saline Water in Agriculture	*I.C. Gupta*
Crop Production in Waterlogged Saline Soils	*S.K. Gupta*
Plant Analysis - Research Methods	*S.S. Narwal*
Biofumigation and Solarization for Management of Soil-Borne Plant Pathogens	*P. Parvatha Reddy*
Soil and Plant Analysis	*C.S. Piper*
Organic Agriculture	*J.C. Tarafdar et al.*
Organic Farming: Theory & Practice	*SP Palaniappan*
Organic Agriculture in the Tropics and Subtropics	*Ulrich kopke*
Challenges and Strategies of Dryland Agriculture	*Rao, S.C.*

SOIL-WATER ENGINEERING
FIELD AND LABORATORY MANUAL

Thomas J. Trout
Ignacio G. Garcia-Castillas
and
William E. Hart

Department of Agricultural and Chemical Engineering
Coloardo State University

SCIENTIFIC
PUBLISHERS (INDIA)

SOIL-WATER ENGINEERING
FIELD AND LABORATORY MANUAL

Water Management Synthesis Project

Reprint , 2012

ISBN: 978-81-7233-723-0

Reprint in India, 2012

by

Scientific Publishers (India)

5-A, New Pali Road, P.O. Box 91
Jodhpur - 342 001 (India)

4806/24, Ansari Road, Daryaganj
New Delhi - 110 002 (India)

www.scientificpub.com

Printed in India

Contents

Foreword

This manual is one of a series supplied to participants in a workshop titled ``Diagnostic Analysis of Farm Irrigation System." The purpose of this manual is to provide the student with basic measurement procedures, suitable for the laboratory and field, for conducting field studies of farm irrigation systems. The original draft of this manual appeared as part of Volume II of the Monitoring and Evaluation Manual. With its evolvement to the present form, a separate volume for soil-water engineering measurements is provided.

The participant in workshop on ``Diagnostic Analysis of Farm Irrigation Systems" should use this manual as a reference for correct laboratory and field measurement procedures. Personnel conducting actual field studies of irrigation systems should also find this manual to be a valuable reference. The equipment lists for each exercise are for determining equipment needs for planning the field study or a workshop. The analysis and discussion suggestions are appropriate for the data collected during the study.

Comments about or suggestions for improving the manual are welcomed and should be sent to the senior author or the Water Management Synthesis Project. Additional copies are available through the Project.

Wayne Clyma
Project Coordinator
Water Management Synthesis Project

Preface to the Manual

This manual is designed to teach basic methods for measuring the soil-water properties most important to irrigation and drainage engineers. The first three exercises describe methods to measure soil properties which affect the soil water. The following three exercises describe methods to measure soil water. Exercises 7, 8 and 9 describe methods to measure parameters which affect soil-water movement. The final three exercises describe measurements which help to determine the quality of soil water for plant use, and the inflows to and outflows from the soil-water reservoir.

Each exercise is divided into five main sections:

1. Introduction
2. Equipment
3. Procedure
4. Analysis and Discussion, and
5. References

The **Introduction** briefly describes the property to be measured and its use to a soil and water engineer. The equations used to calculate the property are developed and the measurement technique is described. Special equipment is also described in this initial section.

The **Equipment** section lists all equipment required to carry out the exercise. The list is divided into two groups. The first group is the special equipment needed for the exercise, followed by a list of equipment which is also used in several other exercises. This often used equipment which is listed in Appendix 1 should be left accessible to the students. An attempt has been made to use as little specialized equipment as possible.

The third section lists the step-by-step **Procedure** to be followed to make the required measurements and the necessary calculations. All collected data can be recorded on the data sheets which are inserted at the end of each exercise. The sheets also should be used to record calculated values. Sections 2 and 3 give all the information required to carry out the exercise.

The fourth section of each exercise describes any **Analysis** which should be done on the results and suggests possible difficulties in the procedures or implications of the findings which should be discussed in the laboratory report. Appendix 2 described several statistical analyses which can be used.

The final section of each exercise lists several sources of additional information on the various procedures. Emphasis is on alternative standard procedures with the methods of the American Society for Testing and Materials (ASTM, 1930), the American Society of Agronomy (Black *et al.,* 1965), the U.S. Bureau of Reclamation (USBR, 1974), and the British Commonwealth Bureau of Soils (1974) referenced when applicable. When more specific information is required regarding the development of or alternatives to the given procedures and instrumentation, it is suggested that the professional literature be searched with assistance from **Soils and Fertilizers,** an indexing and abstracting publication prepared by the Commonwealth Bureau of Soils.

The laboratory exercises can be best carried out in groups of 3 or 4 students. Sufficient space and equipment should be available to allow the groups to work as independently as possible, although items such as balances, ovens, and dessicators can be used cooperatively. Several exercises involve the use of more than one method to make the required measurements. The equipment for these exercises can be used by the different groups in turn.

In order to develop an appreciation for the variability of soil water properties and the difficulty of precise measurements, the groups should exchange results. A comparison and statistical analysis of the results should be included in each lab report.

The ability to describe clearly, in writing, a measurement procedure, the results of a study, and the implications of the results is very important to engineers. Writing and evaluating laboratory reports can improve this ability. Writing reports is thus strongly recommended. A suggested outline for laboratory reports is given in Appendix 3. Because writing good reports requires a significant investment of a student's time, it is suggested that, for each exercise, one member of each group write the report, while the other members only turn in their results and a one page discussion. The report writing responsibilities are then rotated among the group members so that each student would write a total of three or four major reports.

Because soil properties are often interrelated (for example, porosity calculations require both bulk and particle density values), and because it is instructive to measure several properties of the same soil, use of one soil or the same set of soils for as many of the exercises as possible is strongly recommended.

Due to both predictable and unpredictable weather variations, it is possible that the exercises cannot be completed in the order presented. Exercises 2, 3, 4, 7 and 10 are carried out inside the laboratory, while the remainder are field exercises. In sequencing the labs, both the class subject matter and the relationships between lab topics should be considered.

Acknowledgements

This laboratory manual evolved in response to the need for a manual for a Soil-water Engineering Laboratory (AE300) taught at Colorado State University and for a training manual of field procedures for evaluating farm level irrigation water management. The original manual was developed by Ignacio G. Garcia-Casillar and edited by W.E. Hart. That version was then expanded and extensively revised into the present format. In addition Gerald Buchleiter was the primary contributor to Exercise 12.

The manuscript was graciously reviewed by George Hargreaves of Utah State University, Harold Duke and Gordon Kruse of USDA/ARS, and Dave McWhorter. Bill Franklin, Bob Danielson, Gaylord Skogerboe, Walter Baush, and Terry Podmore of Colorado State University. In addition, the students of AE300, who were forced to work from drafts of this manual, offered numerous suggestions and pointed out numerous errors which has led to an improvement in its accuracy and content.

Funding for the original manual was provided by the Egypt Water Use and Management Project (contract AID/NE-C-1351) funded by the U.S. Agency for International Development. Funding for completing the development and printing of this manual was provided by the Department of Agricultural and Chemical Engineering at Colorado State University and the USAID funded Water Management Synthesis Project (contract AID/DSAN-C-0058).

EXERCISE 1

SOIL BULK DENSITY

1.1 INTRODUCTION

The bulk density of soil, $\rho_b \left(g/cm^3\right)$, is the average mass of dry soil solids, M_s (g), per unit volume of soil, V_b $\left(cm^3\right)$:

$$\rho_b = \frac{M_s}{V_b} \tag{1-1}$$

The bulk volume includes the actual volume occupied by the solids or soil grains as well as the pore volume. Bulk densities will therefore vary not only with the actual density of the solids but more importantly with the packing of the soil particles. Consequently, a given soil can have widely varying bulk densities depending upon its state of compacting or aggregation, and the state of the sample must be considered and carefully preserved during collection to achieve an accurate measurement.

Soil bulk density is also sometimes represented as specific gravity (or weight),

$$\gamma_b = \frac{M_s}{M_w} = \frac{M_s / V_b}{M_w / V_w} = \frac{\rho_b}{\rho_w} \qquad \left(V_b = V_w\right) \tag{1-2}$$

In the metric cgs system, because the standard density of water is one, $\rho_b = \gamma_b$. The standard density unit in the International System of units (SI) is kg/m^3, which is 10^3 times the density in cgs units.

Soil bulk density is determined by collecting a known volume of soil, oven drying it, and determining its mass. The main difficulty is to accurately determine the volume of a soil sample in its natural state. Three basic methods used are

1. Inserting a rigid container of known volume such as a soil tube into the soil and extracting a sample.

2. Extracting a soil sample and then determining the volume of the resulting hole, and
3. Collecting a soil aggregate or ``clod'', coating it with an impermeable material such as a paraffin, and determining its volume by fluid displacement (Archimedes' Principle).

The primary difficulty with the first method is compaction of the soil sample when the container is inserted due to friction against the leading (cutting) edge and walls. Pounding with a hammer or force from a hydraulic or mechanical driver is often required to insert the container. The slow, even insertion possible with a hydraulic ram minimizes sample compaction. A difficulty with the second method is collecting the sample without compacting or disturbing the walls of the hole or losing part of the sample. A separate apparatus, such as a water-filled balloon or a measured volume of sand, is required to measure hole volume. The third technique is the most difficult and time consuming of the three, but is useful if the density of an irregular shape such as a soil aggregate is required. The first two methods will be used in this exercise.

1.2 EQUIPMENT (see Fig. 1-1)

1. ``Oakfield'' or comparable soil sampling probe or tube
2. Hammer-driven, double ring soil sampling equipment such as a ``Uhland'' sampler
3. Hydraulic soil sampling ring such as a ``Giddings'' rig, and tubes (Fig. 1-2)
4. ``Volumeasure'' or comparable rubber balloon or volume measurement device (Fig. 1-3)
5. Large spoon
6. Calipers

Also : spade or shovel, spatula, soil sampling cans, oven, dessicator, balance, and meter stick.

Note : ``Oakfield'' probes and ``Volumeasure'' apparatus are available from Soiltest, Inc., 2205 Lee St., Evanston, Ill. 60202. ``Uhland'' samplers are available from Utah State University, Technical Services, UMC 12, Logan, Utah

84322. ``Giddings" rigs are manufactured by the Giddings Machinery Co., P.O. Box 2024, Ft. Collins, Co. 80522.

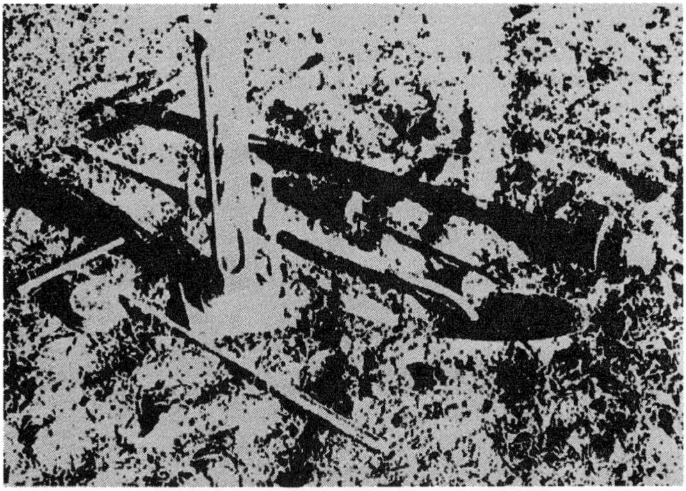

Figure 1-1. Equipment used to collect bulk density samples showing (from top right) ``Uhland" sampler and hammer, ``Volumeasure" balloon apparatus, and ``Oakfield" soil probe.

Figure 1-2. ``Giddings" Hydraulic soil sampler mounted on a tractor.

Figure 1-3. ``Volumeasure" balloon apparatus.

1.3 PROCEDURE

1. Prepare the soil surface by removing vegetation and loose soil and smoothing the surface.
2. Collect volumetric soil samples by each method given below at 0-30 cm and 30-60 cm depths.
3. Transfer the samples to soil cans and oven dry at 105°C for 24 hours.
4. Determine the oven-dry soil sample mass with a balance.
5. Calculate the bulk density of each sample.

Sampling Procedure 1 : Manual soil probe

1. Insert the probe manually as smoothly as possible to the desired sampling depth.
2. Carefully remove the core from the tube and select a portion of the core that includes as much of the depth strata to be sampled as possible.

3. Smoothly cut the selected portion from the remainder of the core with the spatula.
4. Measure the length of the sample and the inside diameter of the soil tube cutting edge.

Sampling Procedure 2 : Hydraulic soil probe

1. Follow the Sampling Procedure 1 above, except the probe is inserted with the hydraulic rig.

Sampling Procedure 3 : Pounded double ring soil probe

1. Assemble the double ring soil sampler with the ring insert.
2. Excavate the soil to the appropriate depth and drive the probe into the soil with the hammer provided.
3. Carefully disassemble the probe, remove the insert ring, and remove the excess soil from the ends of the insert with a spatula.
4. Transfer the insert and/or soil sample to the collection can.
5. Measure the length of the insert and the inner diameter of the tip.

Sampling Procedure 4: Excavation method with rubber balloon assembly (Fig. 1-4).

1. Place the field plate on the test site. If a subsurface sample is desired, the soil must be excavated to the required depth.
2. Place the measurement assembly on the field plate.
3. With the control valve closed, set the air pump for pressure operation, open the control valve, pump the water-filled balloon onto the test site and record the volume of water in the cylinder. Do not overpump, or the apparatus will be lifted off the soil surface giving a faulty reading. This is the initial reading.
4. Close the control valve, reverse the air pump to vacuum operation, open the control valve and return the water and balloon to the cylinder.

5. Remove the reservoir assembly without disturbing the field plate.
6. Excavate a soil sample through the center hole of the field plate, leaving a hole with a diameter of approximately 10 cm and a depth of approximately 12 cm. A tablespoon is a suitable tool for excavating the hole. Remove all the loose soil from the hole. All excavated soil should be placed in a soil can.
7. Place the reservoir assembly on the field plate.
8. With the control valve closed, set the air pump for pressure operation, open the control valve, pump the water-filled balloon down into the excavated hole and read the volume of the water in the cylinder. Do not overpump. This is the final reading.
9. Reverse the air pump to vacuum position and return the water and balloon to the cylinder.
10. Calculate the volume of the excavate hole by subtracting the initial reading (Step 3) from the final reading (Step 8).

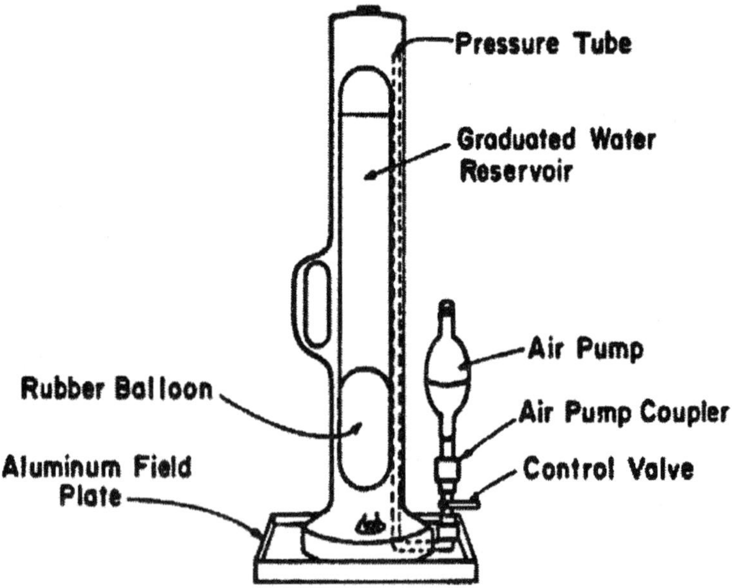

Figure 1-4. Rubber balloon assembly on the field plate.

1.4 ANALYSIS AND DISCUSSION

Determine the average and interval estimate (at a 95 percent probability) soil bulk density for each method and each depth using all groups data. Compare the results. Explain any reasons for differences between the results of the various methods. Describe important factors in collecting accurate bulk density samples (i.e., soil condition, container geometry, driving method).

1.5 REFERENCES

American Society for Testing and Materials (ASTM). 1980. Density of Soil in Place by the Drive-Cylinder Method (ASTM D2937-71), Density of Soil in Place by the Rubber Balloon Method (ASTM D2167-66), and Density of Soil in Place by the Sand-Cone Method (ASTM D1556-64), in Annual Book of ASTM Standards. ASTM, 1916 Race St., Philadelphia, PA.

Black, C.A., D.D. Evans, J.L. White, L.E. Ensminger, and F.E. Clark, 1965. Methods of Soil Analysis, Part I. Chap.30, Bulk Density by G.R. Blake, ASA Monograph Series No. 9. American Society of Agronomy, Madison, WI.

Commonwealth Bureau of Soils. 1974. Method for Analysis of Irrigated Soils, J. Loveday, ed. Chap. 5, Bulk Density, by D. S. McIntyre and J. Loveday. Technical Communication No. 54. Commonwealth Agricultural Bureau, Farnham Royal, Bucks, England.

Soil-Water Engineering Laboratory
Exercise 1
Bulk Density Data Sheet

Method 1: Manual Soil Probe
2: Hydraulic Soil Probe
3: Pounded Soil Probe
4: Excavation/Balloon

Name____________________
Date____________________
Sample Location____________________
Soil Description____________________

Method	Depth (cm)	Sample No.	Sample Volume						Can No.	Sample Mass			Bulk Density ρ_b (g/cm^3)
			Methods 1, 2, 3			Method 4							
			Tube Tip Inner dia. (cm)	Sample Length (cm)	Sample Volume V_b (cm^3)	Initial Reading (cm^3)	Final Reading (cm^3)	Extracted Volume V_b (cm^3)		Can and Sample Dry Mass (g)	Can Mass (g)	Sample Mass M_s (g)	

EXERCISE 2

SOIL PARTICLE DENSITY AND POROSITY

2.1 INTRODUCTION

Porosity

The pores in a soil are the voids between the solid grains of the soil. The air and soil water solution in the soil is stored in and moves through these pores. Although the geometry of these pore spaces is important in the determination of soil water storage and movement, the total pore volume is also a widely used parameter. The pore volume of a soil will vary somewhat with the size and shape of the soil grains, but the soil structure is a more important determinate.

The porosity of a soil, ϕ, is the ratio of the volume of pores, V_P, to the bulk volume of a soil sample, V_b.

$$\phi = \frac{V_P}{V_b} \tag{2-1}$$

Soil pore volume can be measured directly using a gas pycnometer and the relationship between gas pressure and volume (Boyles Law). Since the pore volume is equal to the difference between the bulk volume and the volume of the soil solids, V_s, the porosity can also be calculated if V_s is known:

$$\phi = \frac{V_b - V_s}{V_b} = 1 - \frac{V_s}{V_b} \tag{2-2}$$

If a bulk soil sample could be crushed and compacted to the point where no voids remained, V_s could be determined directly. However, this is not possible, so an indirect method must be used.

The pore space of an oven dry soil sample contains only air whose mass is insignificantly small. Consequently, the mass of a dry bulk sample is equivalent to the mass of the solids, M_s. The ratio of the solid and bulk volumes in Equation 2-2 can therefore be determined for a sample if both the bulk density, ρ_b, and particle density, ρ_s, is known. Since:

$$\rho_s = \frac{M_s}{V_s}, \tag{2-3}$$

then :

$$\phi = 1 - \frac{V_s}{V_b} = 1 - \frac{M_s / \rho_s}{M_s / \rho_b} = 1 - \frac{\rho_b}{\rho_s}. \tag{2-4}$$

Bulk and particle specific gravities $(\gamma_b \ and \ \gamma_s)$ can be used in place of densities.

Particle Density

The particle density of a soil is the density of the soil solids or the average density of the individual soil grains. It will always be larger than the bulk density of the soil.

The volume of a known mass of soil grains can be obtained by using a container of known volume and the Archimedes' Principle that a body submerged in a volume of water will displace a volume of water equal to the volume of the submerged body.

A container of known volume is called a pycnometer. Pycnometers hold a standard volume of distilled water at 20°C. Although pycnometer volumes will increase slightly with temperature, the variations are small and as long as temperatures are maintained within $\pm 5^{\circ} C$ of the standard temperature, no volume correction is necessary. Pycnometers are of two basic types: volumetric flasks and stoppered bottles. The bottles have a tight-fitting ground glass stopper with holes to vent excess liquid. Although coarse-ground soils which are easily wetted and transferred can be manipulated in a pycnometer bottle, a larger volumetric flask is more practical for most soils.

The most direct method to measure particle density would be to add a known mass of soil, M_s, to a known volume of water, V_W, and note the volume of the mixture, V_{SW}. The particle density would be the mass of the soil divided by the volume increase:

$$\rho_s = \frac{M_s}{V_{sw} - V_W} \tag{2-5}$$

However, unless the soil grains are fairly large (i.e., sand and gravels), it is difficult to disperse the soil in the water in a container in which the volume change can be accurately measured. Transferring the mixture from one container to another without losing some of the mixture is also difficult. Consequently, a known mass of soil will be added to a volumetric flask, which is then filled with water.

The mass of water which has been ``displaced" by the soil added to the flask, M_{dW}, would be :

$$M_{dW} = M_{fW} + M_S + M_{fsw} \tag{2-6}$$

where :

M_{fW} = the mass of the flask filled with water, and

M_{fsW} = the mass of the flask filled with soil-water mixture.

By the Archimedes' Principle, the volume of displaced water, V_{dw}, must equal the volume of the added soil, V_s. Since these volumes are equal $\left(V_s = V_{dw}\right)$:

$$\rho_s = \frac{M_s}{V_s} = \frac{M_s}{M_{dw} / \rho_w} = \frac{\rho_w M_s}{M_{fw} + M_s - M_{fsw}} \tag{2-7}$$

The density of water, ρ_w, varies slightly with temperature, and thus a value of ρ_w should be selected from a table such as Table 2-1. It is evident from the listed values that the correction is more academic than practical. Also, since impurities in water affect its density very little, tap water can generally be used instead of distilled water.

The most serious source of error in the experimental determination of soil particle density is not temperature or imputities in the water. The major error will occur from not properly removing entrapped air from the soil-water mixture. Soil, when added to water, will hold entrapped air. If the air is not removed, the volume of entrapped air will result in a decrease in M_{fsw}, which will result in too small a computed value of ρ_s. Use of de-aerated water and removing entrapped air from the mixture by boiling and/or with application of a vacuum will minimize the error. Another source of error, which can also be important, is the use of masses obtained from improperly balanced scales or from two different poorly adjusted scales.

Table 2.1. Density of water at various temperatures.

T (°C)	Density of Pure water ρ_w (g/cm^3)
4	1.0000
15	0.9999
20	0.9982
25	0.9971
30	0.9957
35	0.9941

Typical values of ρ_s which can be used as a guide in determining whether the test results are reasonable are given in Table 2-2. Most agricultural soils with less than 5 percent organic matter have particle densities between 2.5 and 2.7. An often used average value is 2.65.

Table 2.2. Typical values of soil particle density, ρ_s (adapted from Bowles (1970)).

Type of Soil	ρ_s (g/cm^3)
Sand	2.65 - 2.67
Silty sand	2.67 - 2.70
Inorganic clay	2.70 - 2.80
Soils with micas or iron	2.75 - 3.00
Organic soils	Variable but may be under 2.00

2.2 EQUIPMENT

1. 100 ml volumetric flasks

2. Vacuum source regulated to no more than 30 kPa (9 in. Hg) and fittings for the flasks
3. Small funnel
4. De-aerated water
5. Small pipette

Also: Balance ($\pm 0.1g$), mortar and pestle, #6 sieve, 250 ml beakers, thermometer, wash bottle, oven, dessicator and soil sample cans, de-aerated water

2.3 PROCEDURE

1. Prepare de-aerated water by attaching a vacuum to a container of water for at least 24 hours. Be sure the container is rated for the applied vacuum. Allow the water to stabilize at room temperature. Use only this prepared water in the experiment.
2. Weigh the dry volumetric flask to obtain M_f.
3. Carefully fill the volumetric flask with the de-aerated water and weigh to obtain M_{fw}. Measure the temperature of the water. Calculate the flask volume at the measured temperature using ρ_w taken from (or linearly interpolated from) Table 2-1. Empty the flask.
4. Crush the soil aggregate with a mortar and pestle. Screen out large gravel by passing the soil through a #6 sieve. Measure out a portion of the sieved soil, weigh, oven dry (105°C for 24 hours), and reweigh to determine the soil moisture content.
5. Measure out a small amount (about 30 g) of sieved soil, M_s, and mix with de-aerated water in a beaker to form a creamy paste. Add enough additional water to the paste to form a suspension, but keep the total volume of the mixture less than 1/2 the volume of the flask.
6. Transfer the soil water mixture to the volumetric flask. A small funnel will prevent losing any of the sample. Be sure to wash all the soil from the beaker into the flask. This process is difficult with some soils. If all the soil cannot be transferred to the flask, M_s can be determined after

completion of the volumetric measurements by oven drying the soil-water mixture. Now, add more water to the flask until it is two-thirds to three-quarters full. Do not fill water into the neck of the flask as this will reduce the efficiency of the vacuum.

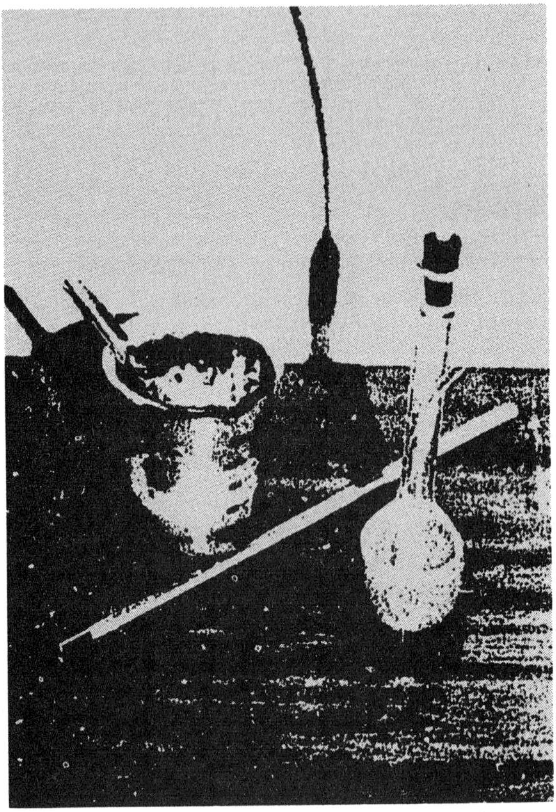

Figure 2-1. Removing entrapped air from the soil-water mixture with a vacuum.

7. Attach the flask to the vacuum for at least 10 minutes, as shown in Figure 2-1. Volumetric flasks are not designed for use under a vacuum, so vacuums above 30 kPa should not be used and care should be taken while handling the flasks. During this time, gently agitate the mixture by carefully shaking and turning the bottle. Observe that the reduced air pressure in the flask causes the water to ``boil." To check the efficiency of the air-removal process, fill the flask into the neck about 1 cm, reattach the vacuum to the flask and

observe if the water level remains substantially at the same level. If the water rises in the neck of the flask under vacuum, the air has not been completely removed. Allow the suspension to sit for a short period of time to clear, then remove enough water to get the surface below the bottom of the neck with a pipette and continue the vacuum process until all the air has been removed. For fine soils, this may take 30 minutes or more. For some soils, the air removal can be accomplished more efficiently by boiling (with heat) in conjunction with the vacuum application.

8. When the entrapped air removal process is complete, carefully add de-aerated water until the bottom of the meniscus is exactly at the volume mark. Care must be taken at this point to avoid reintroducing the entrapped air into the mixture. Organic residue in the sample may float to the top and make the meniscus difficult to see. This residue can be siphoned off with a pipette.
9. Weigh the flask and its contents to the nearest 0.1 g to obtain M_{fsw}. Measure the temperature of the mixture.
10. Compute ρ_s using Equation 2-7.

2.4 ANALYSIS AND DISCUSSION

Determine the mean and confidence interval for the particle density for each soil sample utilizing data from each group. How do the results compare to the values given in Table 2-2. Is the variability reasonable? What might be its causes?

Is the measured value of the flask volume close to the given valuc? If not, whcrc might the error lie?

2.5 REFERENCES

American Society for Testing and Materials (ASTM). 1980. Specific Test Method of Specific Gravity of Soils (ASTM D854-58) in Annual Book of ASTM Standards. ASTM, 1916 Race St., Philadelphia, PA.

Black, C.A., D.D. Evans, J.L. White, L.E. Ensminger and F.E. Clark, 1965. Methods of Soil Analysis, Part I, Chap. 29, Particle Density, by G.R. Blake. ASA Monograph Series #9. American Society of Agronomy, Madison, WI.

Bowles, Joseph, E. 1970. Engineering Properties of Soils and their Measurement. McGraw-Hill, N.Y., p. 56.

Bureau of Reclamation (USBR). 1974. Earth Manual, Appendix E-10 : Specific Gravity of Soils. U.S. Dept. of Interior, USBR. Second Ed.

Commonwealth Bureau of Soils, 1974. Method for Analysis of Irrigated Soils, J. Loveday, ed. Chap. 8, Pore-Space and Aeration Determinations, by D.S. McIntyre, Technical Communication No.54. Commonwealth Agricultural Bureau, Farnham Royal, Bucks, England.

Soil-Water Engineering Laboratory
Exercise 2
Particle Density and Porosity Data Sheet

Name____________________

Date____________________

Sample Description 1.____________________

2.____________________

3.____________________

Test	1	2	3
Volume of flask at 20°C, V_f (cm^3)			
Mass of flask, M_f (g)			
Mass of flask and water, M_{fw} (g)			
Water temperature, T (°C)			
Density of water at T, ρ_w (g/cm^3)			
Measured flask volume, V_f(cm^3)			
Air-dry moisture content sample + can mass (g)			
Oven-dry moisture content sample + can mass (g)			
Mass of soil can (g)			
Air-dry moisture content, W			
Mass of air-dry soil added to the flask (g)			
Mass of dry soil added to the flask, M_s (g) (M_s = (1-W) x Air-Dry Mass)			
Mass of flask and soil-water mixture, M_{fsw} (g)			
Mass of displaced water, M_{dw} (g) ($M_{dw} = M_{fw} + M_s - M_{fsw}$)			
Soil particle density, ρ_s (g/cm^3) ($\rho_s = \rho_w M_s/M_{dw}$)			
Soil bulk density, ρ_b (g/cm^3) (from Exp. 1)			
Soil porosity, $\phi = 1 - (\rho_b/\rho_s)$			

EXERCISE 3

PARTICLE SIZE ANALYSIS

3.1 INTRODUCTION

The most important single piece of information to know about an agricultural soil is the distribution of particle sizes, often referred to as texture. Soil textures have been catagorized into soil types such as sands or clays for easier use. Particle-size distribution has an important influence on a soil's permeability or water intake rate, its water storage capacity, its ability to aggregate and propensity for crusting, and the chemical makeup of the soil water. Particle-size distribution is also important in the design of gravel filters for drainage tile or well screens and for numerous engineering applications involving construction with earth materials.

Particle-size (or grain-size) analysis is a procedure to determine the relative proportions of the different particle (or grain) sizes which make up a given soil mass. Two techniques will be used in this exercise to separate the soil particles into particle-size ranges. Coarse particles (sands and gravels) can be separated with mechanical sieves. The distribution of fine particle sizes (silts and clays) are determined by uniformly dispersing the soil in water and measuring how quickly the particles fall in the mixture. In this exercise, a hydrometer will be used to measure the fall rate, which can be related to particle size by Stokes' equation. Other sedimentation methods, such as the pipette method, involve sampling the soil water mixture to determine the density of the mixture at different depths over time.

Sieve Analysis

Sieve analysis involves passing a soil sample through sieves with successively smaller holes. Consequently, the method cannot determine individual particle sizes. It only divides the particles into size categories bracketed by the sieve opening sizes. All particles retained on a given sieve in a stack are larger than the holes in that sieve but smaller than the previous sieve's holes. By dividing the mass retained on each sieve by the total mass, the percent of the particles in each size range can be determined.

Sieves are made of woven wire screen. They are available in many hole sizes from 100 mm to 0.037 mm. The smallest practical size is the 0.074 mm #200 sieve, since it is difficult to even pass water through smaller holes. Occasionally the #270 sieve (0.053 mm) is used to separate the sand from the silt fraction, but sieve analysis is usually used only to separate out sand and gravel particles.

The openings in sieves are square. The opening size is the distance between two sides of the square opening. Since soil particles are highly irregular in shape, some particles with longer dimensions will occasionally pass through a given sized opening while others with a shorter dimensions may never become properly oriented to pass through the opening. Sieve analysis data must be accepted with this limitation. Also, neither sieve analysis nor hydrometer analysis provides information about the shapes of the particles.

A more serious limitation of sieve analysis of soil materials is the difficulty of separating all the particles into their elemental grains without crushing some of the grains. Also, finer particles will often adhere, due to electrostatic forces, to larger particles, and thus will be measured as larger-sized particles.

The best method of performing sieve analysis on soils is to pulverize an air-day sample, then wash the fines through a #200 sieve. That portion of the sample retained on the #200 sieve is easy to pulverize after oven drying and no fine particles will remain to attach to the larger particles.

Hydrometer Analysis

When soil particles are evenly dispersed in a fluid, the density of the suspension is initially uniform. As time passes, the particles begin to settle out with the larger particles settling out most quickly. Consequently, the density of the suspension varies vertically throughout the suspension and continually changes with time. A hydrometer suspended in the soil-water mixture will indicate the density at the elevation of its bulb, which is in turn related to the amount of sediment still suspended at that elevation.

As the particles continue to settle out, the hydrometer will fall deeper in the suspension. The fall of the hydrometer equals the fall rate of the particles which are just larger than those which remain in suspension at the hydrometer midpoint. Stokes' law, based on fluid mechanics principles, gives the velocity of fall of a sphere in a fluid as a function of the diameter of the sphere, its density, and the density and viscosity of the fluid.

$$V = \frac{2\,g\left(\rho_s - \rho_f\right)}{9\eta}\left(\frac{D}{2}\right)^2 = \frac{g\left(\rho_s - \rho_f\right)D^2}{18\eta} \tag{3.1}$$

where :

v = velocity of fall of a sphere (cm/sec),

ρ_s = density of the sphere (g/cm^3),

ρ_f = density of the fluid (g/cm^3),

η = absolute (dynamic) viscosity of the fluid (dyne-sec/cm^2),

D = diameter of the sphere (cm), and

g = acceleration of gravity (cm/sec^2).

Solving Equation 3-1 for D and using the density of water, ρ_w, for ρ_f, one obtains :

$$D = \left[\frac{18\,\eta v}{g\left(\rho_s - \rho_w\right)}\right]^{1/2} \tag{3-2}$$

Thus, a hydrometer reading at any time after the suspension begins to settle will indicate not only the amount of sample particles still in suspension at a given depth, but also, using Eq. 3-2, the diameter of largest particle still suspended at that depth. In other words, the percent of the total sample which is smaller than a given diameter (or the percent finer) is determined. By repeating the readings over several hours as the particles continue to fall, different suspension densities are read at different depths allowing the percentage of the sample which is smaller than several different particle diameters to be determined. This is the information required to divide the sample into particle sizes.

Stokes' law, and consequently Eq. 3-2, is not valid for particles larger than 0.2 mm because they fall rapidly enough to create fluid turbulence, or for very small particles (<0.0002 mm) because they are affected by molecular forces of attraction and repulsion (Brownian movement).

A special hydrometer, called the type 152H (ASTM designation) has been developed to make particle-size analysis easier. The type 152H is calibrated to read directly in grams of soil with a density of 2.65 gm/cm^3 remaining suspended in 1000 cm^3 of fluid of a density of 1.00 gm/cm^3. A 1000 cm^3 hydrometer cylinder is used to hold the soil-water suspension.

Since the temperature of the soil-water suspension is not likely to be 4°C, the density of the soil particles may not be 2.65, and a dispersing agent must be used in the suspension, corrections must be applied to the actual hydrometer reading to make it accurately read the grams of soil still in suspension. The effect of the water temperature and dispersing agent on the hydrometer reading, can be accounted for by making an adjusted zero hydrometer reading (the zero correction value) in a standard jar of clear water with the same quantity of dispersing agent as is used in the soil suspension and at the same temperature as the suspension. Adjusted zero readings, as well as soil solution readings, should be taken at the top of the meniscus, since it is difficult to read the hydrometer scale below the meniscus in the turbid soil-water mixture.

Once the hydrometer reading has been adjusted by subtracting the zero correction value, the percent finer can be computed by simple proportion as :

$$\text{Percent finer} = \frac{R'\alpha}{M_s} \times 100 \tag{3.3}$$

where

R' – grams of soil in suspension at some elapsed time, t (it is also the adjusted hydrometer reading),

M_s = mass of original soil sample placed in suspension, g, and

α = a multiplier to adjust for particles with a density, ρ_s, other than 2.65, defined by :

$$\alpha = \frac{\rho_s/(\rho_s - \rho_w)}{2.65/(2.65 - \rho_w)} = \frac{0.62\rho_s}{(\rho_s - 1)} \qquad (\rho_w = 1.0) \tag{3.4}$$

The velocity of fall of particles in the solution is related to the drop of the hydrometer in the soil-water mixture, L, in a given time, t:

$$v = L / t \qquad (3.5)$$

Equation 3.5 can be inserted into Equation 3.2 to give, rewritten using L in centimeters, t in minutes, and D in millimeters :

$$D = \left[\frac{30\ \eta\ L}{g\ (\rho_s - \rho_w) t} \right]^{1.2} = \left[\frac{.0306\, \eta}{(\rho_s - \rho_w)} \left(\frac{L}{t} \right) \right]^{1/2} \qquad (3.6)$$

This can be simplified to :

$$D = K \left(\frac{L}{t} \right)^{1/2}, \qquad (3.7)$$

since all factors except L and t will be fairly constant throughout the measurement process. The constant, K, is a function of the soil particle density and the water viscosity and density, which in turn vary only with temperature. Thus, K can be evaluated in terms of ρ_s and temperature, as has been done in Table 3.2.

The fall distance, L, is measured relative to the center line of the hydrometer. Consequently, the hydrometer scale reading must be related to the center line. This can be determined for any hydrometer by suspending it at various depths in a fluid, measuring the depth to the center line and recording scale readings. Table 3.3 lists the effective depth, L for various readings, R, for the standard 152H hydrometer. Note that these hydrometer readings must be adjusted to the bottom of the meniscus.

A dispersing agent must be added to the soil-water suspension to neutralize the charges on the soil particles. The particles often have plus or minus charges, and with proper orientation these charged soil grains will be attracted to each other with sufficient force to remain attached, thus creating larger particles. According to Stokes `law, these larger particles will settle faster through the fluid than the smaller particles. Sodium hexametaphosphate (``Calgon") is used as a dispersal agent to neutralize the soil-particle charges.

Presentation of Results

Complete information about the particle-size distribution of a soil can be presented concisely in the form of a particle-size cumulative distribution curve, an example of which is shown in Figure 3.7. Particle sizes are plotted logarithmically, since they cover such a wide range, on the abscissa; while the present of the total particles which are finer is plotted on the ordinate.

Since the derivative of a cumulative distribution curve will yield a ``density" curve for the distribution, the slope of the curve at any particle size will give the relative amount of particles in that size range. A ``flat" portion of the curve indicates few particles in that size range, while a steep portion indicates many like-sized particles.

Agriculturalists normally group particle sizes into three primary categories : sand, silt, and clay, and classify soils according to the percents of each size classification present. Figure 3.1 and Table 3.1 give the USDA size and type classifications.

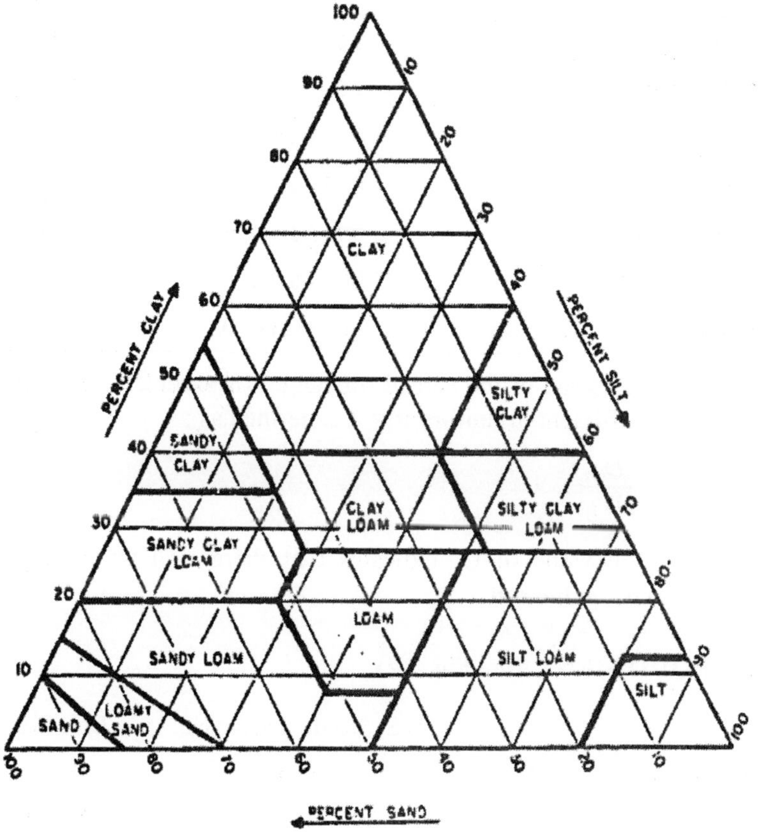

Figure 3.1. Proportions of sand, silt, and clay in the basic soil-textural classes (from U.S. Dept. Agric. 1951. Handbook #18, Soil Survey Manual, p. 503).

For engineering purposes it is usually more convenient to use parameters which describe the shape of the particle-size distribution curve. From the distribution curve, particle sizes such as D_{10}, D_{85}, and D_{50} can be obtained. The D refers to the particle size, or effective diameter of the soil particles, and the subscript (10, 85, 50) denotes the percent which is smaller. For example $D_{10} = 0.21$ mm means that 10 percent of the sample particles are smaller than 0.21 mm.

Table 3.1. Soil particle size classifications (from Soil Conservation Service Engineering Handbook, Section 15 : Irrigation, Chapter 1 : Soil-Plant-Water Relationships. USDA, 1964).

Soil Separate	Particle Diameter (mm)
Very coarse sand	2.0 - 1.0
Coarse sand	1.0 - 0.5
Medium sand	0.5 - 0.25
Fine sand	0.25 - 0.1
Very find sand	0.1 - 0.05
Silt	0.05 - 0.002
Clay	Less than 0.002

An indication of the spread (or range) of particle sizes is given by the ``coefficient of uniformity", C_u, defined as :

$$C_u = \frac{D_{60}}{D_{10}} \tag{3.8}$$

A large value of C_u indicates that there is a wide range of particle sizes in the soil.

3.2 EQUIPMENT (See Fig. 3.2)

		U.S. Standard Sieve No.	Hole Size
1	Set of sieves such as :	#10	2.00 mm
		#20	0.814 mm
		#40	0.420 mm
		#60	0.250 mm

	#140	0.105 mm
including a lid and pan	#200	0.074 mm

2 1,000 cm^3 hydrometer jar (2)

3 152H model hydrometer

4 #13 rubber stopper

5 Soil dispersion device (malt mixer)

6 Soil dispersion agent (sodium hexametaphosphate -- ``Calgon")

7 Soft bristled brush

Also : Mortar and pestle, thermometer, distilled or deionized water, wash bottle, 250 ml beaker, laboratory scoop, watch, soil can, oven, balance, and dessicator

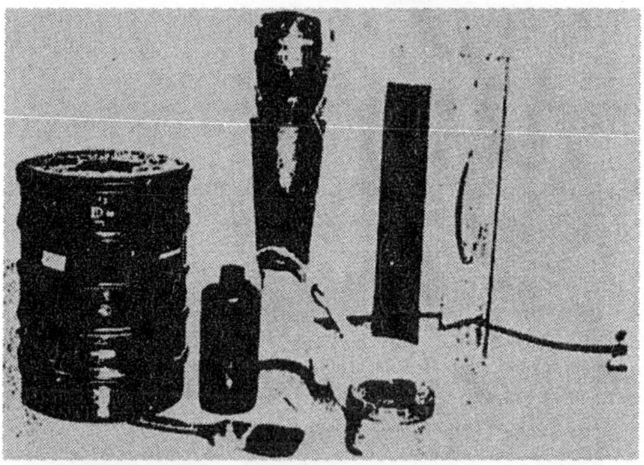

Fig. 3.2 Equipment used to measure particle-size distribution showing (from left), a stack of sieves, soft brush, wash bottle, malt mixer, thermometer, mortar and pestle, soil can, and hydrometer and jars.

3.3 PROCEDURE

Sample Preparation

1. Weigh and crush with a mortar and pestle about 500 g of air dry soil, or enough so that at least 450 g pass through the coarse (#10) sieve.

2. Pass the sample through #10 (200 mm) sieve. Recrush the soil retained on the sieve until all aggregates are broken, and resieve it. Weigh the retained sample. Determine the

percent of the sample which passed through the sieve and record on the Sample Preparation Data Sheet

3 Divide the passed portion into :

   a. about 300 g for sieve analysis.

   b. between 110 g (coarse soils) and 60 g (fine soils) for hydrometer analysis, and

   c. about 50 g to determine the air-dry moisture content.

4. Weigh, oven dry (@ 105°C for 24 hours), and reweigh the moisture content sample. Determine the moisture content of the air-dry sample and record on the Sample Preparation Data Sheet.

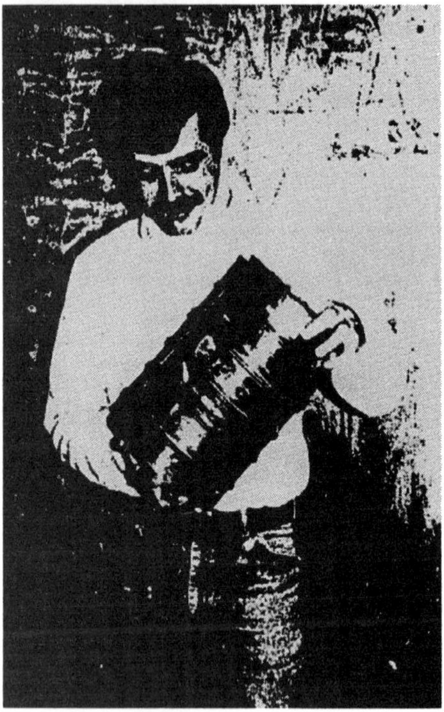

Fig. 3.3. Shaking the stack of sieves.

Hydrometer Analysis

Note : All water used in this analysis should be either distilled or deionized and equilibrated to room temperature.

1. Weigh the hydrometer sample. The sample mass should be between 60 and 110 g depending on the anticipated particle sizes, with finer samples being smaller.
2. Place the sample in a 250 ml beaker and add 125 ml of 4 percent sodium hexametaphosphate dispersing solution. (This solution is made by adding 40 g of dry $NaPO_4$ to 1,000 ml of water). Stir the mixture until thoroughly wetted, and let stand for at least one hour.
3. Pour and wash the mixture into the malt mixer cup. Add deionized water until the cup is at least half full and mix for about one minute.
4. Immediately after mixing, pour and wash the soil-water slurry into the hydrometer cylinder. Fill the cylinder to the 1,000 ml line with deionized water. Place a rubber stopper (#13) over the end of the cylinder and turn the cylinder upside down and back for a period of one minute at about 30 ``cycles" per minute to complete agitation of the slurry. Insure (by shaking if necessary) that particles do not aggregate at either end of the cylinder.
5. Set the cylinder on the counter and immediately begin counting time. Carefully place the hydrometer in the cylinder and take readings (at the top of the meniscus) after 1, 2, 3 and 4 minutes. If the hydrometer floats too high and readings cannot be taken, there is too much soil in the slurry and a new mixtures must be prepared.
6. After the 4th, and each subsequent reading, the hydrometer should be removed from the cylinder and stored in a cylinder of room temperature tap water. The hydrometer is placed in the hydrometer cylinder a few seconds before each reading. This, as well as removal, should be done very carefully, smoothly, and slowly (requiring about 10 seconds) so as not to create turbulence in the cylinder. The hydrometer is removed because soil grains become attached to it.
7. Readings should be taken after about 1, 2, 3, 4, 6, 10, 20, and 40 minutes, 1, 1, 1/2, 2, 2 1/2 and 3 hours, and 1 day.
8. Temperature of the soil water mixture should be measured and recorded periodically.

9. A correction for the density of the dispersant-water solution at the room temperature and the meniscus error in the readings should be determined by placing the hydrometer in a cylinder of dispersant solution at the same temperature and concentration as the mixture in the hydrometer cylinder. This zero correction value reading, R_s (at the top of the meniscus) should be subtracted from the regular readings, R, to get the adjusted readings, R', on the data sheet.

Sieve Analysis

1. Place the weighed sieve analysis sample on the #200 sieve and ``wash" with tap water, the fine particles through the sieve, until the wash water comes out clear. Be careful not to splash soil out the top of the sieve. Being careful to not damage the delicate sieve screen, scoop and wash the retained soil into a soil can, pouring off all excess water after the particles have settled. Dry in an oven at 105°C for about 24 hours.
2. Weigh the dried soil sample, crush it with the mortar and pestle, and transfer it to the top of a stack of about five sieves arranged in decreasing hole size with a pan on the bottom and a lid on top. Suggested sieve sizes so that data points are well spaced on a semilogrithmic plot are listed in the Equipment section.
3. Shake the sieve stack for at least 1 minute (Fig. 3.3).
4. Weigh the soil retained on each sieve. This can be done either by weighing the sieve and soil and then removing the soil to get a tare weight for the sieve, or by pouring and brushing the soil into a weighing container or onto a paper. The finer sieves are very delicate and ulmost care must be used in removing the soil to not damage them. A soft brush such as a paint brush should be used to clean the screen.

Calculations

Note : Sample data and calculated results are shown on Figures 3-4 to 3.7.

Soil-Water Engineering Laboratory
Exercise 3
Particle-Size Analysis

Sample Preparation Data Sheet

Name JOHN SMITH
Date [illegible]
Sample Description SMITH FARM, FIELD #2, SAMPLE #1

Total Air-Dry Sample Mass and Tare (g)	[illegible]
Tare (g)	[illegible]
Total Air-Dry Sample Mass (g)	[illegible]
Mass of Air-Dry Sample Retained on the #10 sieve (g)	[illegible]
Percent of Sample Retained on the #10 sieve (%)	[illegible]
Percent of Sample Passing the #10 sieve (%)	[illegible]
Air-Dry Mass of Moisture Content sub-sample + tare (g)	[illegible]
Oven-Dry Mass of Moisture Content sub-sample + tare (g)	[illegible]
Tare (g)	[illegible]
Air-Dry Moisture Content, W	[illegible]

Figure 3-4. Sample data--sample preparation.

Hydrometer Analysis

1. Convert the air-dry sample mass to oven dry mass by multiplying by one minus the air-dry moisture content, (1-W).

2. Determine the Total Effective Hydrometer Sample mass, M_s, by dividing the oven-dry hydrometer sample mass by he percent of the total sample which passed through the coarse (#10) sieve.

3. Calculate the particle density correction factor (α) from Equation 3-4 for the estimated or measured particle density.

Soil-Water Engineering Laboratory
Exercise 3
Particle-Size Analysis
Hydrometer Analysis Data Sheet

Name: JOHN SMITH

Date: 2/6/81

Sample Description: SMITH [illegible]

Hydrometer sample mass (air dry) (g): 51.51 [illegible]

Hydrometer sample mass (oven dry) (g): 51.15

Total Effective Hydrometer Sample Mass (oven dry), M_s (g): 64.45

Particle density, ρ_s (g/cm³): 2.60 ; Correction factor, α[1]: 1.01

Time	Elapsed time t (min)	Hydrometer reading R (g/ℓ)	Zero Corr. reading R_s (g/ℓ)	Temp (°C)	Adjusted reading R' (g/ℓ)	K[2] factor K	Effective[3] depth L (cm)	Particle[4] diameter D (mm)	Percent smaller P (%)
11:13	0								
	1	27	3	25°	24	.0134	12.4	.047	[illegible]
	2	24	3	25°	21	.0134	12.9	.031	[illegible]
	3	25	3	25°	20	.0134	13.0	.027	[illegible]
	4	22.5	3	25°	19.5	.0134	13.1	.024	[illegible]
	7	22	3	25°	19	.0134	13.2	.018	[illegible]
	10½	21.2	3	25°	18.2	.0134	13.3	.015	[illegible]
	17½	20.8	3	25°	17.8	.0134	13.3	.011	[illegible]
	46	19.8	3	25°	16.8	.0134	13.5	.007	[illegible]
	71	19.0	3	25°	16	.0134	13.7	[illegible]	[illegible]
9:23	1150	16.5	3	25°	13.5	.0134	14.1	.0015	[illegible]
1:13	2520	15.5	3	25°	12.5	.0134	14.3	[illegible]	[illegible]

[1]From Eq. 3-4
[2]From Table 3-2
[3]From Table 3-3
[4] $D = K\sqrt{L/t}$
[5] $P = \frac{R'\alpha}{M_s} \times 100$

Figure 3.5. Sample data--hydrometer analysis.

4. Subtract the zero correction factor reading, R_s, from the hydrometer reading, R, to get the adjusted hydrometer reading (R′).
5. Calculate the percent smaller (passed), P, by Equation 3-3 :

$$P = \frac{R'\alpha}{M_s} \times 100 \qquad (3\text{-}3)$$

Soil-Water Engineering Laboratory
Exercise 3
Particle-Size Analysis
Sieve Analysis Data Sheet

Name ______________________________

Date ______________________________

Sample Description ______________________________

Sample mass + tare (g) ____________________

tare (g) ____________________

Sample mass (air dry) (g) ____________________

Sample mass (oven dry) (g) ____________________

Sample mass retained on #200 sieve (oven dry) + tare (g) ____________________

tare (g) ____________________

Sample mass retained on #200 sieve (g) ____________________

Sample passing #200 sieve (g) ____________________

Percent passing the #10 sieve (%) ____________________

Total Effective Sample Mass (g) ____________________

Sieve #	Hole size (mm)	Mass retained (g)	Mass passed (g)	Total Mass Passed (g)	Per pas (

*Sum of wet and dry sieving.

Figure 3-6. Sample data--sieve analysis.

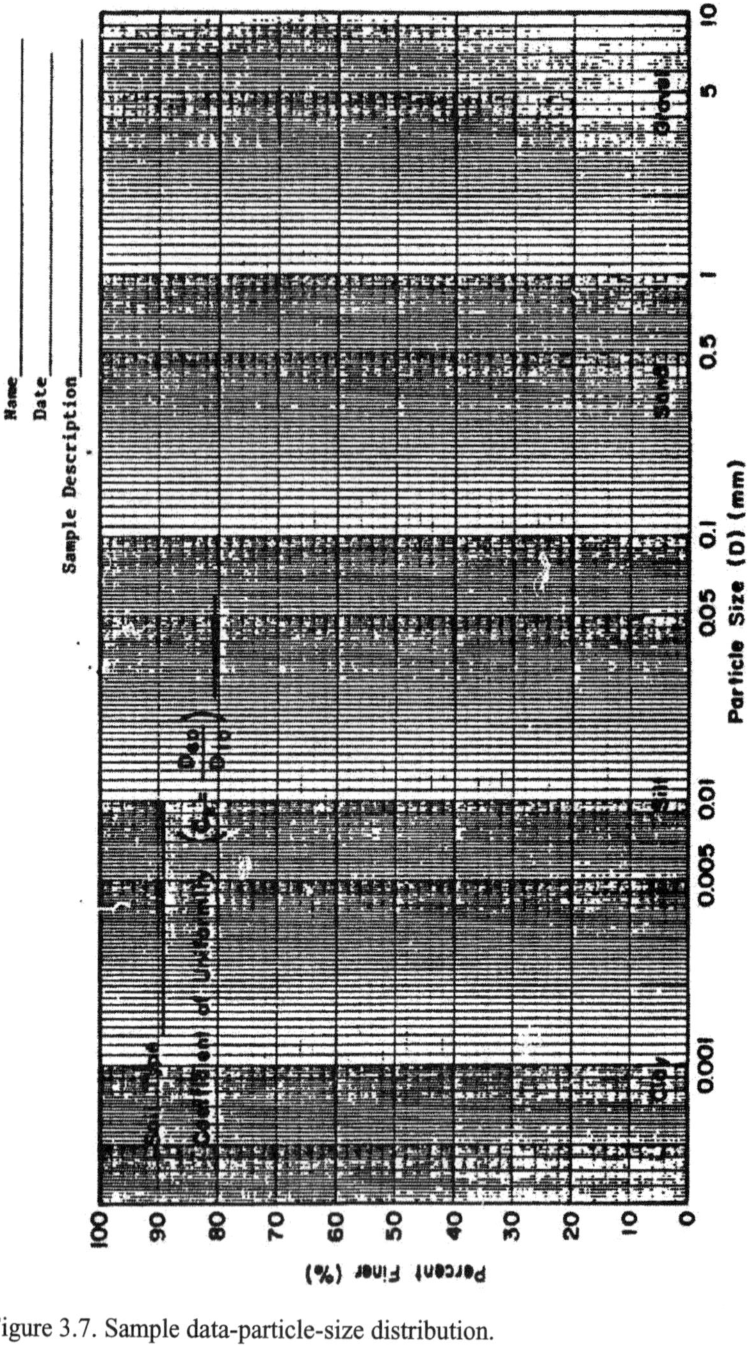

Figure 3.7. Sample data-particle-size distribution.

6. Read the K factor from Table 3-2 for the appropriate temperature and particle density.

Table 3-2. Values of K for use in Equation 3-7 for several unit weights of soil solids and temperature combinations (adapted from ASTM (1980)) p. 222.

Temp. (°C)	Density of soil solids, S_s (g/cm^3)							
	2.50	2.55	2.60	2.65	2.70	2.75	2.80	2.85
16	0.0151	0.0148	0.0146	0.0144	0.0142	0.0139	0.0137	0.0136
17	0.0149	0.0146	0.0144	0.0142	0.0140	0.0138	0.0136	0.0135
18	0.0148	0.0144	0.0142	0.0140	0.0138	0.0136	0.0134	0.0133
19	0.0145	0.0143	0.0140	0.0138	0.0136	0.0134	0.0132	0.0132
20	0.0143	0.0141	0.0139	0.0137	0.0134	0.0133	0.0131	0.0129
21	0.0141	0.0139	0.0137	0.0135	0.0133	0.0132	0.0129	0.0127
22	0.0140	0.0137	0.0135	0.0133	0.0131	0.0129	0.0128	0.0126
23	0.0138	0.0137	0.0134	0.0132	0.0130	0.0128	0.0126	0.0124
24	0.0137	0.0134	0.0132	0.0130	0.0128	0.0126	0.0125	0.0123
25	0.0135	0.0133	0.0131	0.0129	0.0127	0.0125	0.0123	0.0122
26	0.0133	0.0131	0.0129	0.0127	0.0125	0.0124	0.0122	0.0120
27	0.0132	0.0130	0.0128	0.0126	0.0124	0.0122	0.0120	0.0119
28	0.0130	0.0125	0.0126	0.0124	0.0123	0.0121	0.0119	0.0117
29	0.0129	0.0127	0.0125	0.0123	0.0121	0.0120	0.0118	0.0116
30	0.0128	0.0126	0.0124	0.0122	0.0120	0.0116	0.0117	0.0115

7. Read the effective depth, L, from Table 3-3 for the various hydrometer reading values, R, adjusting only for the meniscus.

8. Calculate the particle diameter, D, by Equation 3-7:

$$D = K\left(\frac{L}{t}\right)^{1/2} \tag{3.7}$$

Table 3-3. Values of L (effective depth) for use in Stokes' formula for diameters of particles for ASTM soil hydrometer 152H (adapted from ASTM (1980) p. 221.

Original Hydrometer reading (corrected for meniscus only)	Effective depth L (cm)	Original Hydrometer reading (corrected for meniscus only)	Effective depth L (cm)	Original Hydrometer reading (corrected for meniscus only)	Effective depth L (cm)
0	16.3	21	12.9	42	9.4
1	16.1	22	12.7	43	9.2
2	16.0	23	12.5	44	9.1
3	15.8	24	12.4	45	8.9
4	15.6	25	12.2	46	8.8
5	15.5	26	12.0	47	8.6
6	15.3	27	11.9	48	8.4
7	15.2	28	11.7	49	8.3
8	15.0	29	11.5	50	8.1
9	14.8	30	11.4	51	7.9
10	14.7	31	11.2	52	7.8
11	14.5	32	11.1	53	7.6
12	14.3	33	10.9	54	7.4
13	14.2	34	10.7	55	7.3
14	14.0	35	10.5	56	7.1
15	13.8	36	10.4	57	7.0
16	13.7	37	10.2	58	6.8
17	13.5	38	10.1	59	6.6
18	13.3	39	9.9	60	6.5
19	13.2	40	9.7		
20	13.0	41	9.6		

Sieve Analysis

1. Determine the oven-dry sample mass by multiplying the air dry mass by one minus the air-dry moisture content (1-W).
2. Determine the sample mass passing the #200 sieve by subtracting the retained sample oven-dry mass from the total sieve sample oven-dry mass.
3. Determine the masses which passed through each sieve by summing the masses retained on each finer sieve. The mass passed plus that retained on the coarsest sieve should equal the mass of the sample retained on the #200 sieve. If it is off by more than 5 percent, too much soil has been lost during sieving and the sieve analysis should be rerun.
4. Add the mass which passed the #200 sieve during the washing to each ``Mass Passed" to get the ``Total Mass Passed".
5. The "Total Mass Passed" plus the mass retained on the coarsest sieve is the total mass passed by the #10 sieve during the sample preparation. This value divided by the percent passed by the #10 sieve (from the sample Preparation Data Sheet) is the "Total Effective Sample Mass."
6. Divide all "Total Masses Passed" by this "Total Effective Mass" to get percent passed by each sieve, which is the percent finer than each sieve size.

Presentation of Results

1. Plot percent passed (finer) for both the sieve and hydrometer analysis on the 4 cycle semilog graph paper provided. Differentiate between the two types of data points.
2. Determine the percent sand, silt, and clay and soil type using Figure 3-1 and Table 3-1. Note that gravel (>2.00 mm) is not considered soil in this classification. All other fraction percentage must thus be increased proportionately.
3. Calculate a coefficient of uniformity, C_u, for the soil from Eq. 3-8.

3.4 ANALYSIS AND DISCUSSION

Do the curves of the sieve and hydrometer fit together smoothly? If not, discuss some reasons.

Since only 3 size ranges are normally used in classifying agricultural soils, how can this procedure be simplified and the basic information still be gained.

What can you predict about the soil-water properties of the measured soils based on their particle-size distribution?

3.5 REFERENCES

American Society for Testing and Materials (ASTM). 1980. Standard Methods. Standard Method of Particle-Size Analysis of Soils (ASTM D422-63) in Annual Book of ASTM Standards. ASTM, 1916 Race ST., Philadelphia, PA.

Black, C.A., D.D. Evans, J.L. White, L.E. Ensminger, and F.E. Clark. 1965. Methods of Soil Analysis, Part I. Chap. 43, Particle Fractionation and Particle-Size Analysis by P.R. Day. AsA Monograph Series No. 9, American Society of Agronomy. Madison, WI.

Bureau of Reclamation (USBR). 1974. Earth Manual, Appendix E-6, Gradation Analysis of Soils. U.S. Dept. of Interior, USBR, Second Ed.

Commonwealth Bureau of Soils. 1974. Method for Analysis of Irrigated Soils, J. Loveday, ed., Chap. 11, Particle Size Analysis by D.S. McIntyre and J. Loveday. Technical Communication No. 54. Commonwealth Agricultural Bureau, Farnham Royal, Bucks, England.

United States Department of Agriculture, Soil Conservation Service. 1979. Engineering Field Manual. USDA, SCS.

EXERCISE 4

SOIL WATER HOLDING CAPACITY

4.1 INTRODUCTION

Soil, due to the geometry of its pore spaces between the soil particles, has the capacity to hold water. This very important trait allows soil to retain precipitation or irrigation water in the root zone to be used by plants over time. The amount of water held in a soil depends upon the soil's porosity and pore-size distribution, and varies with the capillary pressure of the water in the soil. Consequently, a fine textured soil will hold a different amount of water at a given capillary pressure than a coarse-textured one, and any soil will hold different amounts of water at different capillary pressures.

The relationship between the amount of water held in a soil and the capillary pressure of the water is depicted by a moisture release or desorption curve. Figure 4-1 shows a typical moisture release curve for a loamy soil. Two regions of the curve are of special interest to agriculturalist-the field capacity or residual saturation and the permanent wilting point. The field capacity represents the water content at which the moisture level in the soil begins to remain relatively constant as the capillary pressure continues to increase. It is often defined as the water content at one-third bar pressure since many soils reach this plateau at about 33 kPa (1/3 bar). In the field, field capacity is defined as the water which remains in the soil after the soil has drained to a deep water table.

Permanent wilting point is the moisture content at which plants will permanently wilt. It will vary with both plant and atmospheric conditions, but, being primarily dependent on the energy status of the soil water, is usually defined as the soil water content at a capillary pressure of 1,500 kPa (15 bars).

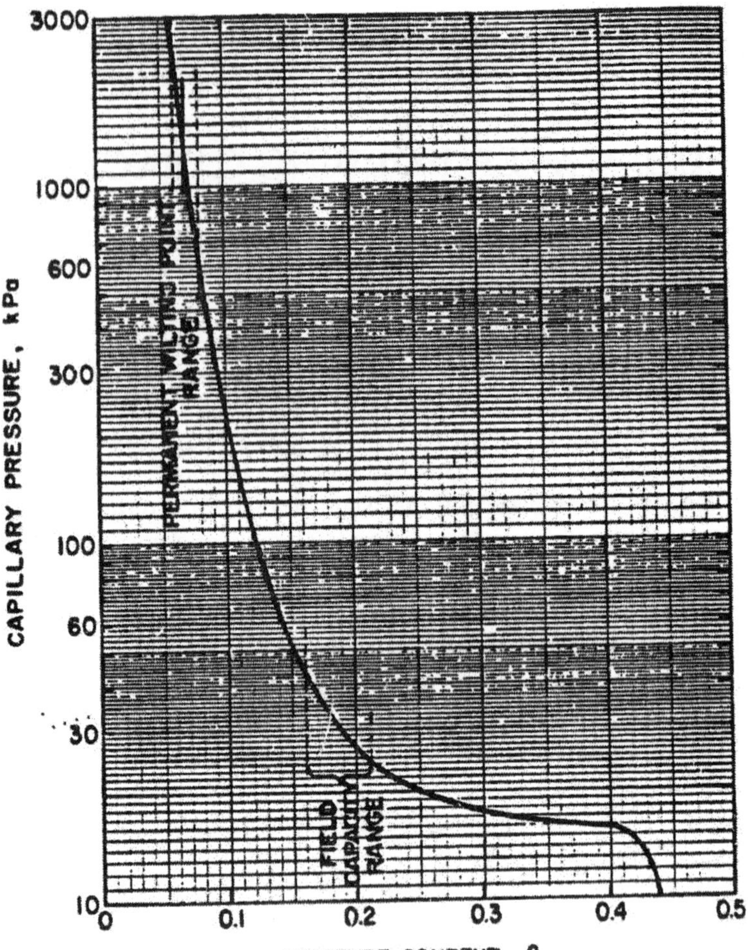

Figure 4-1. Moisture release curve for a loamy soil.

A soil's water holding capacity for plant use, termed the total available water, TAW, is defined as the difference between the amount of water held in the plant's root zone when the soil is at field capacity FC, and permanent wilting point, PWP :

$$TAW = \left(FC - PWP\right) \rho_b \ Y, \tag{4.1}$$

where FC and PWP are on a dry mass basis, ρ_b is the soil bulk density (g/cm^3) which converts from gravimetric to volumetric water contents and Y is the depth of the root zone.

The soil water content vs. capillary pressure relationship for soil is measured in the laboratory by creating a differential pressure between the soil water and the surrounding air while allowing water to move out of the soil sample. After flow has stopped, the moisture content of the sample is determined gravimetrically. By repeating this procedure a several pressures, a moisture release curve can be constructed. A porous ceramic plate is used to maintain the required pressure differential while permitting the flow of water out of the sample. The bubbling pressure or air entry pressure of the ceramic must be high enough to prevent desaturation of the previously saturated plate at the desired capillary pressures. Consequently, the pore sizes in the ceramic must be small, resulting in low permeability, relatively slow water movement, and fairly long time periods to reach equilibrium.

The pressure differential can be created either by reducing the pressure on the water in the ceramic plate (which is transmitted to the water in the soil in contact with the plate), or by increasing the pressure above atmospheric in the atmosphere around the soil. The soil only ``feels" the differential. Because pressure in a moist environment can be reduced only to the vapor pressure of water (or equivalently negative gage pressures or suctions can be created only as high as atmospheric pressure minus the vapor pressure) positive gage pressures which effectively drive the water out of the soil, are used.

Figure 4-2. Pressure plate extractor.

A pressure plate extractor, shown in Figures 4-2 and 4-4, is a device specially designed for determining water content : capillary pressure relationships. The extractor is composed of a container or cell capable of containing high pressures fitted with a removable ceramic plate. The soil samples are placed on top of the saturated ceramic plate as shown in Figure 4-3. The underside of the plate, which is often encased in a flexible envelope, is connected via a tube, to the outside of the cell. Thus, the top side of the plate is at the pressure applied to the cell while the underside is at atmospheric pressure. Water extracted from the sample also passes out of the cell through the tube.

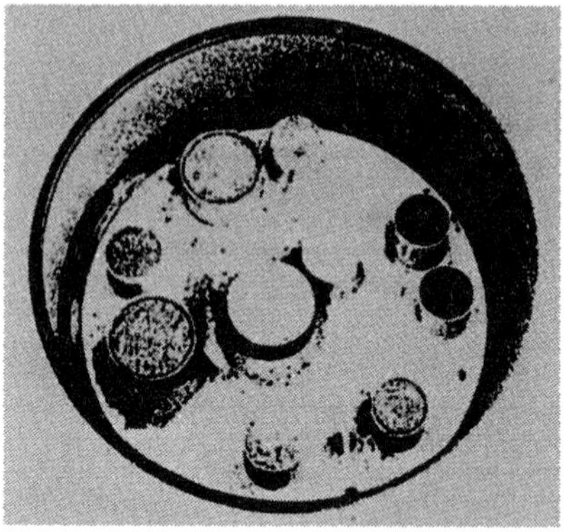

Figure 4-3. Soil samples in place on the ceramic plate in a pressure cell.

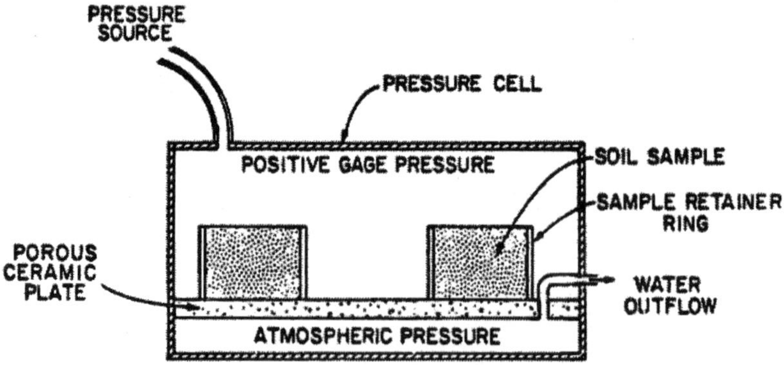

Figure 4-4. Schematic of a pressure plate extractor.

Field capacity of a soil can also be determined in the field by saturating a soil and allowing it to drain while inhibiting evaporation or transpiration from the surface. Moisture samples collected from the draining soil will indicate an initial rapid water content decrease followed in usually 1/2 (coarse soils) to 3 (fine soils days by a reduced rate of decrease as indicate in Figure 4-5. The soil water content at the time when the rate decreases would be field capacity.

An alternative, physiologically more precise, means of determining the permanent wilting point of a soil utilizes an indicator plant, such as a healthy sunflower, grown in a container of the soil. The soil is allowed to dry until the plant wilts and, even in a humid environment, does not revive. This procedure is described in more detail in Black *et al.* (1965).

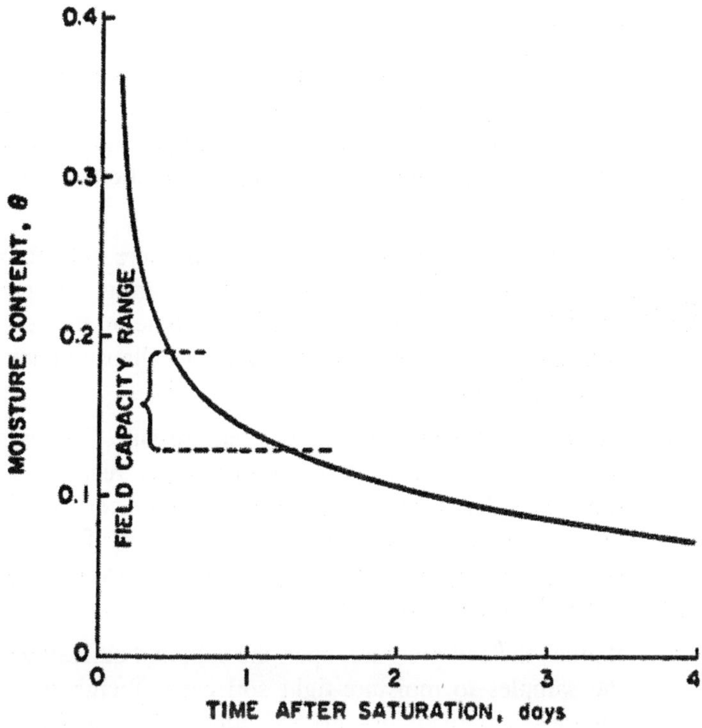

Figure 4-5. Moisture content vs. time after saturation for a field soil draining to a deep water table.

4.2 EQUIPMENT

1. Pressure plate extractor
2. Regulated air pressure source to 1,500 kPa
3. 100 kPa (1 bar) and 1,500 kPa (15 bar) ceramic plates
4. Soil sample retainer rings
5. Syringe or pipette

Also : oven, dessicator, balance, soil cans, mortar and pestle, #6 sieve, spatula

4.3 PROCEDURE

1. Place the soil samples in the retainer rings. Disturbed samples should be ground with a mortar and pestle and passed through a #6 sieve.
2. Place the samples on the previously saturated place and place the plate in shallow water so that the samples are partially submerged. Allow the samples to saturate by imbibition for several hours. Sample saturation can be done in the cell.
3. Place the plate and samples in the pressure cell. Remove excess water from the plate with a syringe or pipette. Attach the nipple on the plate to the nipple on the cell wall with a piece of stiff (so that it doesn't collapse under high pressures) tubing.
4. Close the cell and apply the desired pressure. The pressure must be maintained until equilibrium is reached in the samples and water no longer drips from the cell outlet. The time required to reach equilibrium will vary from several hours for sandy soils at low pressures to several days for fine soils at high pressures.
5. Reduce the cell pressure, open the cell, and quickly transfer the samples to moisture-tight soil cans. Weigh the moist samples and oven dry (105°C for 24 hours). Cool in a dessicator and weigh the dry samples and determine the sample water content and bulk density.
6. Repeat the above procedures at sufficient different pressures between 10 and 1,500 kPa to sketch a moisture retention curve. Use the 100 kPa (1 bar) plate for pressures up to 100 kPa and the 1,500 kPa (15 bar) plate for higher pressures.

4.4 ANALYSIS AND DISCUSSION

Indicate the field capacity and permanent wilting point on each moisture retention curve. What is the total available water per meter of root zone for each soil tested. How does each parameter vary with soil type? How will the moisture retention curve of undisturbed soil samples compare to that of pulverized samples? Does the 33 kPa moisture content appear to be a good estimate of field capacity for the measured soils?

If the average capillary pressure of the soil from which each sample was taken was measured in the field with a tensiometer to be 65 kPa, how much available moisture remains in a 1.2 m root zone and what depth of irrigation should be applied to refill the soil moisture reservoir?

4.5 REFERENCES

American Society for Testing and Materials, 1980. Capillary Moisture Relationships for Course-and Medium Textured Soils by Porous-Plate Apparatus (ASTM D2325-68) and Capillary-Moisture Relationships for Fine-Textured Soils by Pressure Membrane Apparatus (ASTM D3152-72), in Annual Book of ASTM Standards. ASTM, 1916 Race St., Philadelphia, PA.

Black, C.A., D.D. Evans, J.L. White, L.E. Ensminger, and F.E. Clark. 1965. Methods of Soil Analysis, Part I, Chap. 18, Water Capacity, by A. Klute, and Chap. 19, Water Availability, by D.B. Peters, ASA Monograph Series No. 9. American Society of Agronomy, Madison, WI.

Hagan, R.M., H.R. Haise, T.W. Edminster, 1967. Irrigation of Agricultural Lands, Chap. 12, Nature of Soil Water, by P.R. Day, G.H. Bolt, and D.M. Anderson; and Chap. 15. Measurement of Soil Water, by P.J. Kramer, D. Biddulph, and F.S. Nakayama. American Society of Agronomy Monograph No. 11. ASA. Madison, WI.

Commonwealth Bureau of Soils. 1974. Methods for Analysis of Irrigated Soils, J. Loveday, ed. Chap. 6, Water Retention and the Moisture Characteristic, by D.S. McIntyre. Technical Communication No. 54. Commonwealth Agricultural Bureau, Farnham Royal, Bucks, England.

Soil-Water Engineering Laboratory
Exercise 4
Water Content vs. Capillary Pressure Data Sheet

Name____________________

Date____________________

Soil Sample Description #1____________________

Soil Sample Description #2____________________

Soil Sample Description #3____________________

Soil Sample Description #4____________________

Sample Volumes

Sample Number	Ring Dimensions	Sample Volume

Pressure (kPa)	Sample Number	Ring No.	Can No.	Wet Mass + tare (g)	Dry Mass + tare (g)	Tare Mass (g)	Mass of water M_w (g)	Dry Mass M_s (g)	Sample Volume V_b (cm^3)	Bulk Density (g/cm^3)	Water Content W(g/g)	Water Content θ (cm^3/cm^3)

EXERCISE 5

SOIL-WATER CONTENT

5.1 INTRODUCTION

A critical environmental factor to plant growth is maintenance of adequate moisture in the root zone. The water is needed not only as a transport medium in which nutrients are absorbed and moved through the plant, and wastes are expelled, but more critically on a short-term basis, as a coolant. Transpiration of water by the plant is required to maintain non-lethal temperatures. A lack of water, for even a short time, can cause death; a shortage of water will cause yield reduction. Since plans absorb nearly all of their water through their roots, maintenance of adequate soil moisture in the root zone is essential.

The water content of a soil can be represented on a dry soil mass basis, a bulk volume basis, or a pore volume basis. On a dry mass basis :

$$W = \frac{M_w}{M_S} \qquad (5\text{-}1)$$

where :

W = the soil water content (g/g),

M_W = the mass of water in a soil sample (g), and

M_S = the mass of oven-dry soil in the sample (g).

On a bulk volume basis :

$$\theta \; \frac{V_W}{V_b} = \frac{M_W}{\rho_w V_b} \sim \frac{M_W}{V_b} \qquad (5\text{-}2)$$

where :

θ = the volumetric moisture content (cm^3/cm^3),

V_W = the volume of water in a soil sample (cm^3), which is essentially equal to the mass of water in the cgs system since the density of water, ρ_W, is nearly 1.0 g/cm^3, and

V_b = the bulk volume (see Exercise 1) of the sample (cm^3).

On a pore volume basis :

$$S = \frac{V_W}{V_P} = \frac{M_W}{\rho_W \; V_P} \sim \frac{M_W}{V_P} = \frac{M_W}{\phi V_b} \tag{5-3}$$

where :

S = saturation (cm^3/cm^3),

V_P = the pore volume (cm^3), and

ϕ = the sample porosity (cm^3/cm^3) (see Exercise 2).

Each of these dimensionless quantities is often converted to a percent by multiplying by 100. The basic measurement is often made on a dry mass mass. Conversion to the other two forms can be made using the sample specific gravity, γ_b (or bulk density, ρ_b $\left(g/cm^3\right)$), and porosity, ϕ $\left(cm^3/cm^3\right)$:

$$\theta = \frac{M_W}{\rho_W \; V_b} = \frac{\rho_b}{\rho_W} \frac{M_W}{M_S} = \gamma_b \, W \sim \rho_b \, W, \text{ and} \tag{5-4}$$

$$S = \frac{M_W}{V_P} = \frac{M_W}{\phi V_b} = \frac{\theta}{\phi} = \frac{\gamma_b}{\phi} W \sim \frac{\rho_b}{\phi} W \tag{5.5}$$

since specific gravity is numerically nearly equal to density in the cgs system.

Several direct and indirect methods are available for measuring soil-water content. Direct methods are those in which the water in the soil is removed from the soil by evaporation, distillation or chemical reaction. Determination of the amount removed may be by :

1. Measurement of the weight loss of the soil sample by evaporation (gravimetric).
2. Collection of water by distillation. The soil sample is mixed with mineral oil (of a higher boiling point than water) and heated. The water is driven off by heating and collected by distillation to give the original water content in the soil sample.

3. Extraction of water with substances which will replace it is the sample and measurement of some physical or chemical property of the extracting material which is quantitatively affected by water content (i.e., the ``Speedy Moisture Tester" sold by Soiltest, Inc.).

Two indirect methods are those of neutron thermalization and gamma ray absorption. In the neutron thermalization or scattering method a radioactive source is placed in the soil and the emitted neutrons are thermalized (slowed down) by water held in the soil. The density of slow neutrons is then measured. In the gamma ray absorption method, the degree to which a beam of monoenergetic gamma rays is attenuated (reduced in intensity) in passing through a soil is measured. If the density of the soil is constant, then changes in the attenuation represent changes in water content.

In this exercise, soil-water content will be measured gravimetrically and by the neutron themalization method.

Gravimetric Method

Water content determination by the gravimetric method involves collecting and weighing a moist sample, removing the water by oven drying, and weighing the dry sample. Water content is determined by Equation 5-1 which can be expanded as follows:

$$W = \frac{M_W}{M_S} = \frac{M_{wst} - M_{st}}{M_{st} - M_t} \tag{5-6}$$

where :

M_{wst} = mass of wet sample plus tare (container),

M_{st} = mass of dry sample plus tare, and

M_t = tare mass.

Two factors are important in this simple process. First, the sample must not be allowed to lose water between collection and weighing. This implies use of an air tight sample collection can (taped closed if storage time will be more than a few hours) and weighing as soon as possible after collection. Plastic bags, which are somewhat permeable to air, should be used only if storage time will be short and high accuracy is not critical.

Second, a ``standard" dryness must be reached. Some water (called hygroscopic or molecular) is held by soil even after drying at above 100°C for several days. However, high temperatures wil oxidize the organic matter. Consequently, drying at 105°C until a constant mass is attained (usually less than 24 hrs) is the commonly accepted definition of oven dry and the standard used for moisture measurements. The moisture still held in a soil after that treatment is very small.

After the sample is dried, it should be cooled before reweighing because the sample heat will create convection currents which will affect sensitive balances. However, cooling in a moist atmosphere will allow the hot soil to absorb moisture. Thus, the dried sample should be cooled in the dry atmosphere of a desiccator and/or with the sample container lid on.

Soil samples for gravimetric moisture determinations can be collected with a shovel or spade (surface samples), soil augers, or manually inserted, hammer driven, or hydraulically inserted soil sampling tubes. Figure 5-1 shows a King tube sampler being driven into the soil.

Fig. 5-1. King tube being driven into the soil to collect subsurface moisture samples.

Fig. 5-2 Neutron probe resting on an access tube and connected to a scaler.

Neutron Thermalization

The atomic age has provided a modern, practical method of determining soil water content through measuring the degree to which high energy neutrons are slowed down in the soil. When high energy neutrons are emitted from a source of radiation material, they collide with atomic nuclei that are nearby. If the nuclei with which they collide are of a heavier mass, they will bounce off retaining practically all their original velocity. If, however, the nuclei with which the neutrons collide are of approximately the same mass as the neutrons, they will be slowed down. Such neutrons are said to be thermalized and they can be detected as described below.

In soils there are few elements that are effective in thermalizing neutrons. The atomic nucleus most effective in slowing neutrons is that of hydrogen, which has the same mass as a neutron. Lithium, beryllium, boron and carbon may also thermalize neutrons but they are less effective as their nuclei are more massive. In soil these latter elements usually occur in small to insignificant amounts, but hydrogen occurs in large amounts as a component of water.

A probe containing a source of high energy neutrons and a counter for detecting the thermalized neutrons in lowered into an aluminium or steel tube that has been placed in the soil. Because hydrogen is chiefly responsible for thermalizing the neutrons, the rate at which thermal neutrons are detected is proportional to the number of hydrogen nuclei present in the vicinity of the source and detector, provided that the rate of emission of fast neutrons and the geometry of the area through which the neutrons are scattered are constant.

The volume of soil measured by neutron thermalization will vary with the moisture content. The radius of primary influence in a moist soil is only about 8 cm, but will extend to 30 cm in dry soils. Because of this sphere of influence, measurements should not be made closer than 18 cm to the soil surface.

The density of slow neutrons relative to the rate of high energy neutrons emitted will be roughly proportional to the density of hydrogen atoms in the soil, or the volumetric moisture content, θ. However, the proportionality constant between thermalized neutron detection and moisture content will vary from probe to probe and with different access tubes. A calibration curve is often supplied with a probe which can be used if the recommended access tube is used. However, calibration of the probe and tube is preferable. Calibration with one soil can be assumed to hold for all soil types unless the presence of a substance which will affect the thermalization process is suspected. Figure 5-3 shows a neutron probe calibration curve.

Figures 5-2 and 5-4 show the equipment used in the neutron thermalization method. The probe contains a source of fast neutrons, a detector tube, and a preamplifier. The source is composed of materials such as radium-beryllium (2 to 5 mc) or American beryllium (100 mc). Both materials have long lives (1,600 and 460 years, respectively). The detector tube is commonly a glass container filled with boron-trifluoride plus other inert gasses. When a thermal neutron passes through the gases, it causes a momentary ionization resulting in a brief pulse of current flow. The transistorized preamplifier boosts the pulse signal and sends it to the scaler.

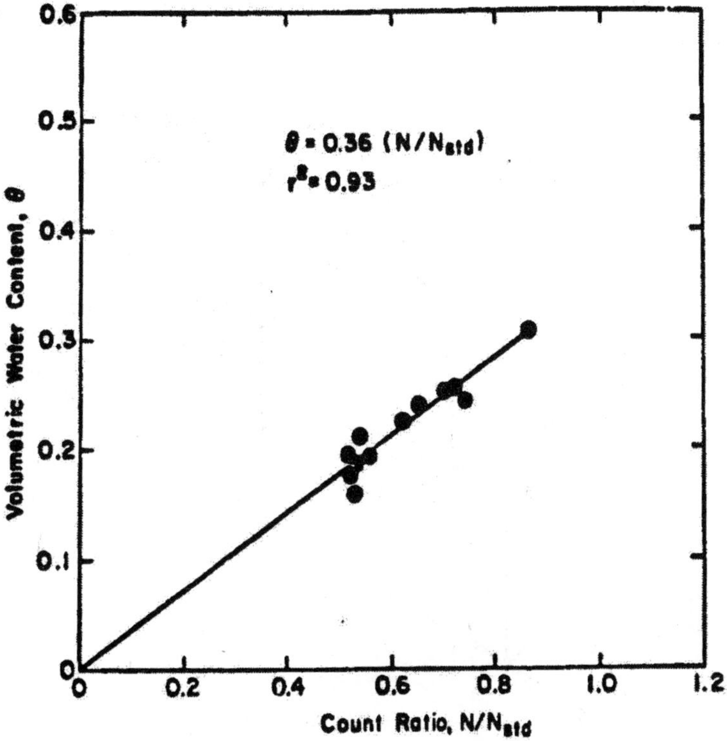

Figure 5-3. Neutron probe calibration curve.

The probe is stored in a shield to contain the constantly emitted high energy neutrons and gamma rays when the probe is not in use. The shield commonly consists of lead and paraffin or polyethylene, and is constructed with a latch such that the contained probe can be released only when set on an access tube of the proper diameter. The shield also acts as a standard absorber and can be used to obtain a standard count.

The counting device can be a scaler, where each pulse from the detector is counted and accumulated for a set period of time; or a rate meter, which indicates the time rate of pulses or thermalized neutron counts. The probe is connected to the counting device with a cable which is usually marked so that the depth of the probe can be easily determined.

Aluminium or steel access tubing is most commonly used. There are two sizes of access tubing -- 1.7 mm (20-gauge) wall thickness tubing with an outside diameter of 4.12 cm (1.62 in.), and

1.3 mm (0.05 in.) wall aluminium tubing with an outside diameter of 5.08 cm (2 in.).

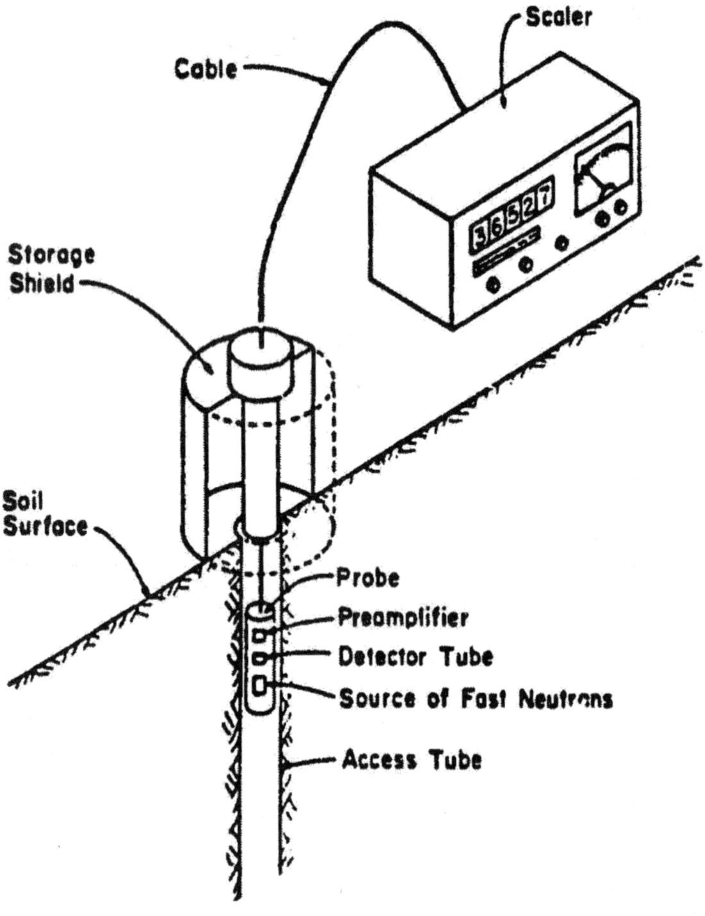

Figure 5-4 Neutron thermalization equipment including probe, shield and scaler.

Even though the radiation given off by a neutron probe is small, attention should be paid to safety measures to minimize any potential health hazard. Important precautions include :

1. Keep the probe in its shield at all times except when it is lowered into the access tube (in the soil) for measurements.

2. Reduce exposure to the small radiation escaping the shield by keeping at least two meters away except when changing

the position of the probe, and by keeping the open end of the probe and shield pointed away from personnel.

3. Do not eat or drink while using a neutron probe, and wash well after use (in case of leakage).
4. Transport the probe in the back of a truck or in a car trunk.
5. When the probe is not in use, lock it in a storage cabinet. Label the cabinet plainly to indicate the presence of radioactive materials.

5.2 EQUIPMENT

Gravimetric Method

1. Auger, soil sampling tube, or other suitable device for collecting soil samples at the appropriate depth

Also : soil cans with tight-fitting lids, oven regulated to 105°C, desiccator with active desiccant, balance ($\pm$ 0.1 g)

Neutron Thermalization Method

2. Neutron probe with shield and cable
3. Scaler or rate meter (in modern devices, the scaler and shield are contained in one unit)
4. Access tubes (see previous section for specifications) with caps
5. Auger to install access tubes

5.3 PROCEDURE

Gravimetric Method

1. Collect soil samples from the desired depth.
2. Place the sample of approximately 100 g of soil in containers with tight-fitting lids.
3. Weigh the samples immediately or store them in such a way that evaporation is negligible until the mass is determined.

4. Place the samples in a drying oven with the lids off, and dry them to constant mass at approximately 105°C for 24 hours.
5. Remove the samples from the oven, replace the covers, and place them in a desiccator containing active desiccant.
6. Weigh the samples when cool.
7. Empty the cans of all soil and reweigh to obtain the tare mass.
8. Determine the gravimetric moisture content with Equation 5-6. If the soil bulk density is known, calculate the volumetric moisture content. If the soil porosity is known, also calculate the saturation percentage.

Neutron Thermalization Method

Steps 1 through 3 describe the installation of access tubes. Once installed, the tubes can be used repeatedly, and even for several seasons if they do not interfere with tillage operations.

1. Drill the access hole vertically with a diameter slightly smaller than the access tube in order to avoid air pockets around it.
2. Close the access tube at the bottom with a rubber stopper or with a metal or plastic plug to prevent water from entering, and insert the tube into the access hole. Leave several centimeters of the tubing above the soil surface and close it with a rubber stopper or with a metal or plastic cover to prevent entrance of trash and water.
3. If the tube is not clearly visible, mark with a stake, ensuring that the stake does not interfere with the measurement process.
4. Place the shielded probe next to or on the access tube and connect the cable to the scaler (or rate meter) located at least 2 meters from the tube. Turn the scaler on to standby power and allow it to warm up for at least 30 min (transistorized units need no warm up) before taking the first reading. Also, return the scaler to standby power between readings.

5. Turn the scaler from standby to on and take a count reading of one minute duration (or rate meter reading) with the probe in the storage shield.

6. Lower the probe into the access tube and take a scaler count of one minute duration (o rate meter reading) at successive depth intervals, starting at least 30 cm from the soil surface. Fifteen centimeter intervals are ordinarily used. These are called depth counts.

7. After completing the readings at different depths return the probe to the storage shield and take a final scaler or rate meter reading. The average of this count and the count taken in Step 5 is called the standard count, N_{std}.

8. Divide depth counts, N, by the standard count N_{std}, to obtain count ratios. If a calibration curve for the instrument and tube is available, calculate the soil-water content on a volume percentage basis.

9. If a calibration curve is not available, collect gravimetric soil moisture samples from several depths from at least 2 holes about 30 cm from each access tube. Also, collect volumetric soil samples for each depth from the immediate area and determine the soil bulk density. (Soil moisture samples can be used to determine bulk density if carefully collected). Calculated the volumetric water content for each access tube and each depth, and plot θ versus the neutron probe count ratio, N/N_{std}, as is shown in Figure 5-3.

5-4 ANALYSIS AND DISCUSSION

Compare the gravimetric and neutron probe soil moisture measurements of the different lab groups. Which is more reproducible? Which is more dependable? Under what conditions would each method be preferable?

Plot the moisture content with depth. What might have caused the water content variability with depth? If the total available water, TAW, and field capacity, FC, for the soil profile is known, what is the percent depletion, f, and soil-water deficit, SWD? (SWD = (FC – W) ρ_b y; $f = (SWD / TAW) \times 100$) If not known, assume FC = 0.20, PWP = 0.10, $\rho_b = 1.3\ gm/cm^3$, $and\ Y = 120\ cm$).

5.5 REFERENCES

American Society for Testing and Materials. 1980. Laboratory Determination of Moisture Content of Soil (ASTM D2216-71), in Annual Book of ASTM Standards. ASTM, 1916 Race St., Philadelphia, PA.

Black, C.A., D.D. Evans, J.L. White, L.E. Ensminger, and F.E. Clark 1965. Methods of Soil Analysis, Part 1, Chap, 7, Water Content by W.H. Gardner. ASA Monograph Series No. 9. American Society of Agronomy. Madison, WI.

Bureau of Reclamation (USBR). 1974. Earth Manual Appendix E-9: Moisture Determination of Soils. U.S. Dept. of Interiror, USBR. Second Ed.

Wilson, R.G. 1971. Methods of Measuring Soil Moisture. International Field Year for the Great Lakes, Technical Manual Series No. 1. U.S. National Academy for Sciences, Washington, D.C.

Van Bavel, C.H.M., P.R. Nixon and V.L. Hauser, 1963. Soil Moisture Measurement with the Neutron Probe Method. USDA/ARS Publication No. 41-70.

Soil-Water Engineering Laboratory
Exercise 5
Soil-Water Content Data Collection Sheet

Name____________________

Date____________________

Location____________________

Neutron Thermalization

Probe Description____________________

Access Tube Description____________________

Initial Standard Count____________________

Final Standard Count____________________

Average Standard Count, N_{std}____________________

Gravimetric

Depth (cm)	Soil Bulk Density ρ_b (g/cm^3)	Soil Porosity ϕ (cm^3/cm^3)

Depth (cm)	Count N	Count Ratio N/N_{std}	Soil-Water Content θ	Depth (cm)	Can No.	Wet Mass + tare (g)	Dry Mass + tare (g)	Tare Mass (g)	Moisture Content W	θ	S

EXERCISE 6

SOIL-WATER CAPILLARY PRESSURE

6.1 INTRODUCTION

The capillary pressure of water in soil indicates the energy status of the soil water, or the force which must be exerted to extract water from the soil. It is thus the best indicator of the availability of the soil water to plants. Due to the measurable relationship between capillary pressure and moisture content, discussed in Exercise 4, capillary pressure can be related to moisture content for a given soil, allowing determination of moisture deficiencies and irrigation requirements.

Two techniques are used to measure soil-water capillary pressure in the field. Tensiometers can directly sense through a saturated porous material, soil-water pressures. The main drawback of tensionmeters is their limited range. The second technique involves placing a stable porous object with a constant moisture content vs. capillary pressure relationship in the soil and monitoring, usually remotely, some property which is related to its moisture content. Resistance measurements of buried gypsum blocks is the most common method used, although heat conductance is also used as are materials other than gypsum. The main drawback with this technique is the difficulty of manufacturing sensors with a consistent relationship between capillary pressure and the measured parameter and the resulting need for calibration.

Tensiometer

Figure 6-1 shows a typical tensiometer, consisting of a porous ceramic cup, an air-tight connecting tube leading to a pressure-sensing device and a covered opening for refilling the tube. Some commercially available tensiometers also have a fluid storage area near the top. The ceramic cup must be buried underneath the soil surface in contact with the soil to be measured. The water-filled connecting tube then leads to the soil surface to allow easy reading of the gauge and access for refilling.

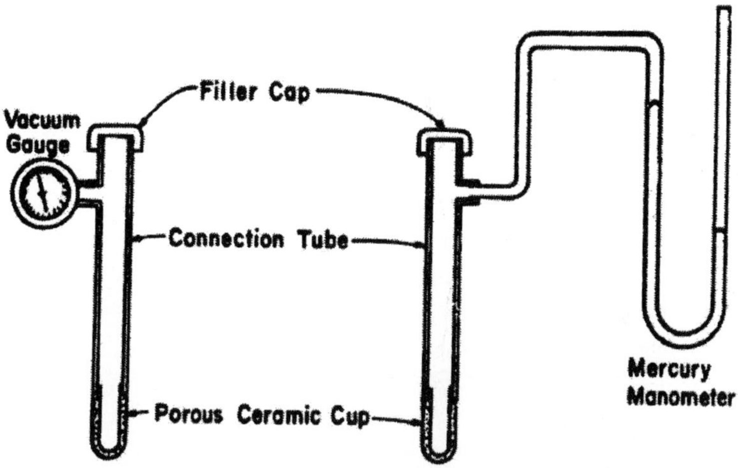

Figure 6-1. Tensiometers with vacuum gauge and mercury manometer.

The porous ceramic tip, when saturated, will allow water to move through it without desaturating and allowing air to pass through, as long as the bubbling or air entry pressure of the ceramic is not exceeded. The principle is the same as with the ceramic plates used in the pressure plate apparatus in Exercise 4. Thus, when the cup is in contact with a soil, water will move through the tip until the pressure on both sides of the ceramic are equal, or until the water inside the tensiometer is in hydraulic equilibrium with the soil water. As the soil dries, water moves out of the tensiometer until its internal pressure is again equal to the increasing soil-water capillary pressure.

If the tensiometer fluid were incompressible and the walls were totally rigid, no flow would have to take place through the ceramic for hydraulic equilibrium to be maintained. However, due to dissolved air in the fluid, slight flexibility of the tensiometer walls, and especially the volumetric change required to activate a pressure sensing device, a small amount of fluid movement through the cup does occur. Tensiometer fluids, materials, and gauges which minimize this required volumetric change are preferred because they reduce the lag time between an actual soil-water pressure change and the pressure change indicated by the tensiometer. The lag is caused by the low permeability of the ceramic cup and the time required for the outflow from the cup to be absorbed and redistributed through the soil (due to the low hydraulic conductivity of the partially saturated soil).

The fluid in the connecting tube transfers the pressure at the tip to the vacuum sensing device. Vacuum gauges are the most common vacuum sensing devices for most field applications. Vacuum gauges with electronic switches are available for remote monitoring and to act as controllers for automated irrigated systems. When increased sensitivity is required, mercury manometers can be used to sense the capillary pressure. One difficulty with manometers is the volume of water required to indicate pressure changes. Pressure transducers can be used when remote sensing of the pressure or minimal volumetric change is required. Figure 6.2 shows tensiometers with vacuum gauges installed in the field.

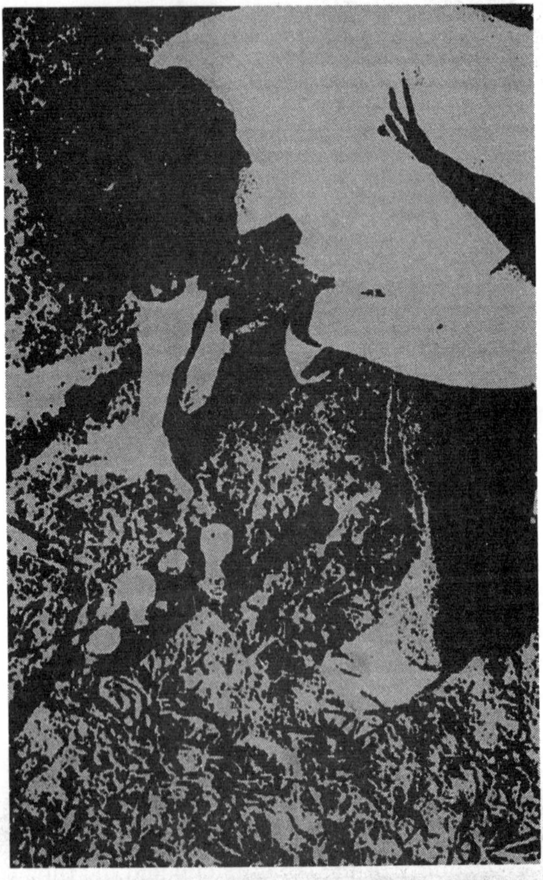

Figure 6.2. Two tensiometers installed to two different depths being read.

Most tensiometer gauge scales read centibars, which are equal to kilopascals (kPa). Some commercial gauges are scaled in terms of soil moisture. These gauges should be used with caution since different soils hold widely varying amounts of moisture at the same capillary pressure.

Because of the column of water which connects the vacuum gauge to the ceramic cup, the vacuum or negative pressure at the gauge will be higher than the vacuum at the cup by the head difference between the two ends. For example, since a 10 cm column of water exerts about 1 kPa of pressure, the gauge on a 60 cm tensiometer will overestimate actual capillary pressure by 6 kPa. The effective zero of the gauge or manometer can be determined by partially submerging the ceramic cup in water, holding the tensiometer vertical, and taking a reading. The accuracy of most gauges will not merit zero adjustments unless long tubes are used.

The minimum pressure which can be created in a saturated environment is the vapor pressure of water at that temperature. Thus, the highest vacuum which can theoretically be created in a tensiometer is atmospheric pressure at that altitude minus the vapor pressure. Consequently, the range of tensiometers is limited and will decrease with increasing altitude and temperature. Atmospheric pressure varies from about 100 kPa at sea level to 85 kPa at 1,500 m. Water vapor pressure varies from about 1 kPa at 10°C to 6 kPa at 40°C. Thus, the theoretical limit of a tensiometer measurement at Fort Collins on a summer day would be about 80 kPa. The actual upper limit is only about 80 percent of the theoretical limit due to the dissolved air in the fluid.

In light of the fact that the wilting point of soils is often estimated by the moisture content at a capillary pressure of 1,500 kPa, the limited range of tensiometers seems to be a serious constraint. However, other than clayey soils will have exceeded a 50 percent depletion of the total available water at a capillary pressure of 75 kPa, and sandy soils will, often reach 75 percent soil-water depletion, SWD, at 50 kPa (see Exercise 4). Thus, for coarser-textured soils or when high moisture levels (low SWDs) are to be maintained, tensiometers can provide an adequate measure of soil moisture levels.

Tensiometers are commercially available from such firms as Irrometer Company, P.O. Box 2424, Riverside, California 92516;

Soilmoisture Equipment Corp., P.O. Box 30025, Santa Barbara, California 93105; Perma Rain Irrigation Inc., P.O. Box 880, Lindsay, California 93247; and Soiltest, Inc., 2205 Lee Street, Evanston, Illinois 60202.

Electrical Resistance Blocks

A block of porous material, when buried in the soil, will come to hydraulic equilibrium with the soil, or equivalently, the capillary pressure in the material will be equal to the capillary pressure of the surrounding soil. This is accomplished by a flow of water into or out of the block toward the higher capillary pressure. Consequently, if the moisture content vs. capillary pressure relationship for the porous material is known, the capillary pressure of the soil can be determined by measuring the moisture content of the material.

The resistance to a flow of electricity of a porous material is affected by its moisture content, since water has a high conductance. If electrodes are embedded in the block, the resistance of the porous material can be used as an indicator of its moisture content and thus of the capillary pressure of the soil.

Figure 6-3 shows one type of electrode arrangement in an electrical resistance block. Other types of electrode arrangements include concentric cylinders of wire mesh, parallel wire mesh plates, and simply two parallel wires.

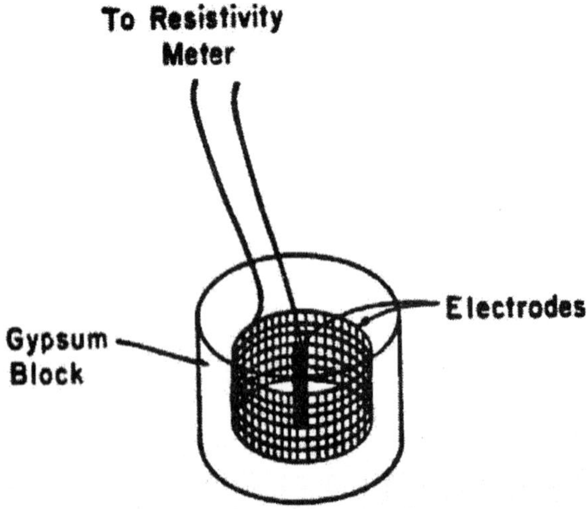

Figure 6-3. One electrical resistance block electrode arrangement.

Blocks are made of several porous materials such as nylon, fiberglass, and plaster, but the most common is gypsum. Because of this, resistance blocks are often referred to as gypsum blocks. Gypsum has the advantage of maintaining a relatively constant ion concentration in the block-that of a saturated solution of calcium sulfate-- and thus reducing the sensitivity of the resistance measurement to the salinity level of the soil water. This effect will be further discussed in Exercise 10. Gypsum blocks will operate adequately in soils with saturated extract conductivities, EC_{25}, up to 2,000 umhos/cm. Gypsum blocks will deteriorate in the soil, especially in sodic soils or where the water table is high. Some blocks may deteriorate in one season, although some stabilized gypsum compounds can also remain intact up to 6 years.

When a current is induced across the two electrodes, the resistance of the moist block material can be determined. The most accurate measurements are obtained with an alternating current Wheatstone bridge, which does not allow current to flow when the measurements are taken. However, measurements are often taken with a simple ohmmeter which records current flow when a known voltage (from a batter) is impressed across the electrodes. Commerical maters give arbitrary readings which are not directly correlated to the resistance of the block. Because block resistance must be calibrated to capillary pressure these arbitrary readings are all that are required. Figure 6-4, shows the resistance of a gypsum block being read with a ``Delmhorst" ohmmeter.

The measured resistance (or other meter reading) of block must be calibrated with known capillary pressures within the block to know the meaning of each reading. The calibration curve shown in Figure 6-5 was obtained in the laboratory from blocks embedded in soil which was placed in a pressure plate apparatus. Commercial blocks often come with a calibration curve. These curves should always be checked in the field.

Resistance block readings can be related to the moisture content of a given soil in the field by collecting gravimeteric soil moisture samples from near the blocks and relating the moisture content of the soil to the readings. For dependable use, resistance blocks should regularly be recalibrated in the field.

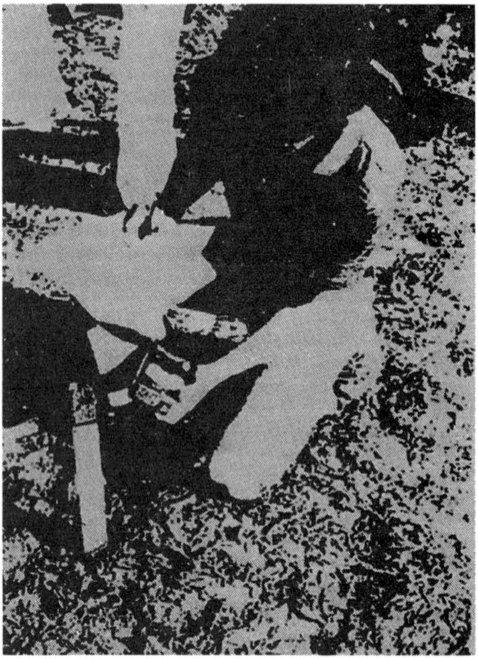

Figure 6-4. Measuring the resistance of a buried gypsum block with a ``Delmhorst" meter.

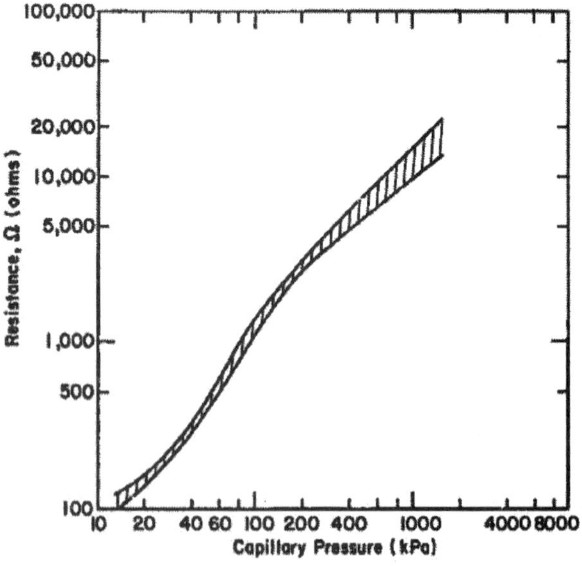

Figure 6-5. Typical calibration curve for one type of gypsum blocks showing the range of data variability.

Electrical resistance blocks and meters are available from several commercial firms such as: Delmhorst Instruments Co., 848 Cedar Street, Boonton, New Jersey 07005; Soilmoisture Equip. Corp., P.O. Box 30025, Santa Barbara, California 93105; and Soiltest, Inc., 2205 Lee Street, Evanston, Illinois 60202.

Scheduling Irrigation

A plot of tensiometer or resistance block readings over time, such as that shown in Figure 6-6, can be an easy and practical measns of monitoring soil moisture and scheduling irrigations. If the readings are calibrated, the soil-water deficit can also be determined, allowing the application of the proper depth of irrigation. Devices installed at various depths in the root zone will monitor moisture distribution in the profile. The devices can also be wired directly to automated irrigation systems allowing the soil moisture status to directly control irrigations.

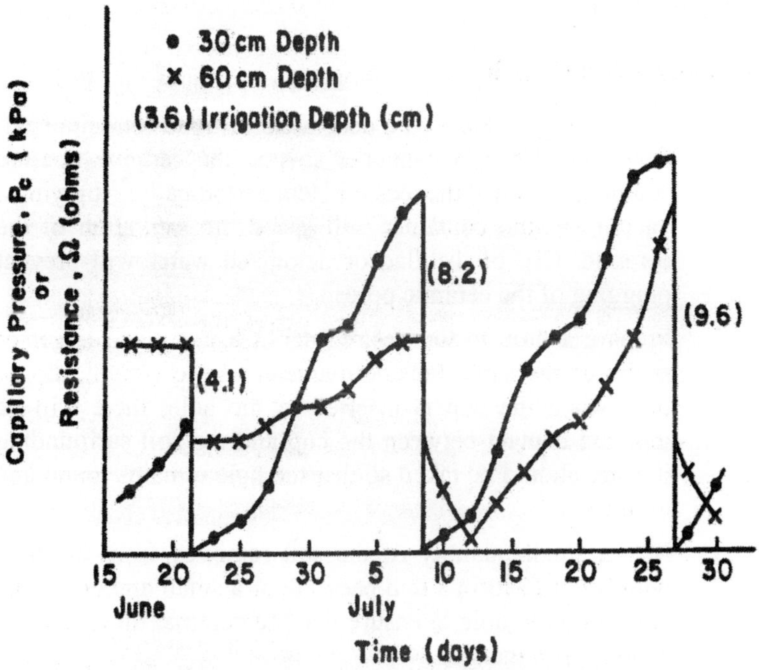

Figure 6-6. Plot of soil capillary pressure measurements over time at two depths showing variations with soil drying and irrigation.

6.2 EQUIPMENT

1. Tensiometers of varying lengths (15 cm, 30 cm, 45 cm, 60 cm and 90 cm).
2. Soil probe, pipe, or auger of the same diameter as the tensiometers
3. Evacuation pump
4. A set of at least 4 electrical resistance blocks
5. Soil probe or auger slightly larger in diameter than the blocks
6. A resistance meter (preferably an alternating current wheatstone bridge)
7. Stakes to mark the block locations

Also: deaerated, distilled (or deionized) water, soil cans, balance, dessicator, and oven

6.3 PROCEDURE

Tensiometer Installation

1. Fill the tensiometers with deaerated, distilled (or deionized) water and set in a container so that the ceramic tips are submerged. Refill the tensiometers periodically. Applying a vacuum to the container will speed the saturation of the ceramic. Use of distilled or deionized water will prevent plugging of the ceramic pores.
2. Prepare a hole to the desired depth using a soil auger or probe of the same outside diameter as the ceramic tip so that, when the cup is inserted in the hole, there will be intimate contact between the cup and the soil surrounding it. Care should be taken so that the hole remains round and uniform.
3. Mix a small quantity of the soil removed from the hole with water to form a thin paste. Pour a small amount of the paste into the hole to ensure that the ceramic tip is in good contact with the soil.
4. Push the tensiometer firmly into the bottom of the hole with a twisting downward motion applied to the connecting tube. Do not push on gauge.

5. Compress soil around the tensiometer at the soil surface and pile it slightly so that water will not collect around the instrument.
6. Fill the tensiometer with deaerted water and apply a vacuum to the top for severals seconds to remove any air trapped in the tube or gauge.
7. Protect tensiometers to prevent field damage by bracketing the instrument with bright colored stakes or covering it with a box, tile or steel pipe. If occasional freezing temperatures may be encountered in the field, the tensiometers must be covered with a bucket or box to prevent ruining the gauge.
8. A vacuum should be applied to the tensiometer occasionally after filling to remove any accumulated air.

Tensiometer Reading

1. Early morning reading is desirable as a condition near equilibrium exists. (Water movement in plants and soil at this time has virtually ceased).
2. Read the tensiometer at about the same time each day.
3. Take readings frequently enough so that the change from one reading to the next is not greater than 10 to 15 kPa on the gauge. If irrigation is more often than once a week, take daily readings. If irrigation is monthly, twice weekly readings are adequate.
4. If the pressure limit of the tensiometer is reached, the water will be drawn out. When this occurs, refill with deaerated water and apply a vacuum to remove any air bubbles.
5. Generally, a reliable tensiometer reading may be made 24 hours after installation.

Resistance Block Installation

1. Submerge the blocks in water to saturate. Applying a vacuum to the container will speed the saturation process.
2. Form a hole with the probe or auger to the proper depth. Make a small amount of paste with the removed soil and

pour into the hole. Insert the block into the hole with a soil probe or rod assuring good contact with the soil.

3. Bring the leads to the surface running them horizontally for a short distance below the surface to assure against any continuous vertical flow paths. Fill the hole. Mark the location with a stake on which the depth is noted.

Resistance Block Reading

Use the same schedule as outlined above the making tensiometer readings.

Moisture Content Calibration in the Field

1. Collect gravimetric soil moisture samples (see Exercise 5) from the vicinity of and at the same depth as the tensiometers or blocks. Two or three samples from various sides of the sensor should be made. Calibration moisture samples should be collected at varying gauge or meter readings so that the full range of the instrument can be calibrated.
2. Weigh the moisture samples, oven dry (105°C for 24 hours), and reweigh to determine the moisture content, W.
3. Plot the moisture content versus meter or gauge reading.

If the moisture content : capillary pressure relationship for the soil is known, tensiometer readings can be directly realted to soil moisture without field calibration. Resistance block readings can be calibrated with capillary pressure in the lab with a pressure plate apparatus (see Exercise 4) which can then also be related, like the tensiometer readings, to soil moisture. However, direct in-field calibration is more practical.

6.4 ANALYSIS AND DISCUSSION

Which of these two types of measurements should be more accurate? Which should be more dependable? Which would you select and why? What would be the percent depletion of each of the soils tested in Exercise 4 at a tensiometer reading of 75 kPa.

6.5 REFERENCES

Black, C.A., D.D. Evans, J.L. White, L.E. Ensminger, and F.E. Clark. 1965. Methods of Soil Analysis, Part 1, Chap. 9, Soil Suction Measurements with Tensiometers by S.J. Richards. ASA Monograph Series No. 9. American Society of Agronomy, Madison, WI.

Wilson, R.G. 1971. Methods of Measuring Soil Moisture. International Field Year for the Great Lakes. Technical Manual Series No. 1. U.S. National Academy for Sciences, Washington, D.C.

Soil-Water Engineering Laboratory
Exercise 6
Tensiometer and Resistance Block Calibration Data Sheet

Name________________________

Location and Description________________________

			Soil Moisture Content					
Date	Tensiometer (Block) No.	Tensiometer (Block) Reading	Wet Mass + tare (g)	Dry Mass + tare (g)	Can (tare) Mass (g)	Moisture Content W (g/g)	Bulk Density ρ_b (g/cm^3)	Moisture Content θ (cm/cm)

Soil-Water Engineering Laboratory
Exercise 6
Tensiometer and Resistance Block Data Record

Field and Location Description________________

Location and Depth of Tensiometer (Block) #1________________

#2________________

#3________________

#4________________

#5________________

#6________________

Date	Time	Reader	Tensiometer (Block) Reading and Equivalent Soil Moisture Content, θ											
			1		2		3		4		5		6	
			Reading	Equiv θ	Reading	Equiv θ	Reading	Equiv θ	Reading	Equiv θ	Reading	Equiv θ	Reading	Equiv θ

EXERCISE 7

HYDRAULIC CONDUCTIVITY

7.1 INTRODUCTION

Fluids move through porous media such as soils under the influence of a driving force. The rate of movement is generally proportional to the driving force :

$$q = k\frac{\Delta h}{\Delta l} \tag{7.1}$$

where :

q = volume flux or bulk velocity of flow (cm/sec);

K = proportionality constant termed the hydraulic conductivity (cm/sect);

$\Delta h / \Delta l$ = summation of driving forces in the direction of flow or hydraulic gradient (cm/cm)

Equation 7.1 is called Darcy's Law. In this exercise, the hydraulic conductivity of saturated soils to water will be determined by two laboratory methods : the constant head and the falling head method.

The hydraulic conductivity is a measure of how easily water can move through soils, whether it be flow toward drains or a pumped well, seepage from canals, or infiltration into an irrigated field, and thus is an important quantity for many agricultural applications.

A permeameter is an apparatus designed to apply a known driving force or hydraulic gradient to the fluid (usually water) in a soil sample to generate a measurable flow rate of the fluid through the sample. Two types of permeameters will be used in this exercise : a constant head (Fig. 7.1), and a falling head permeameter (Fig. 7.2). Both measure the one dimensional flow of water through a column of soil.

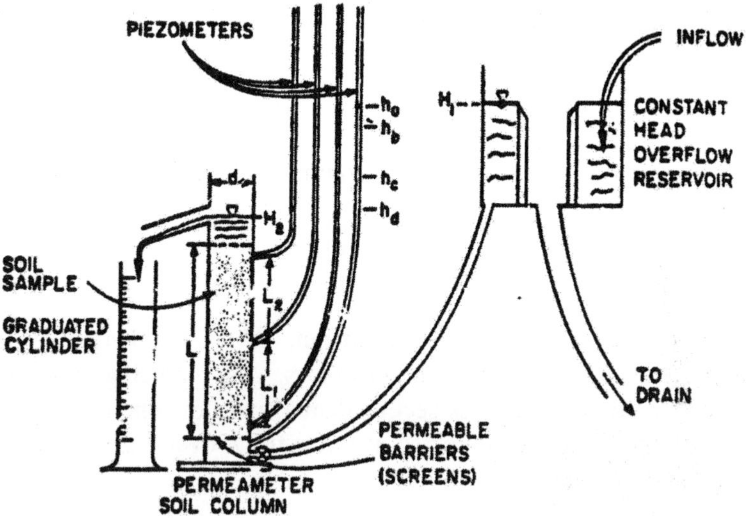

Figure 7.1 Constant head permeameter.

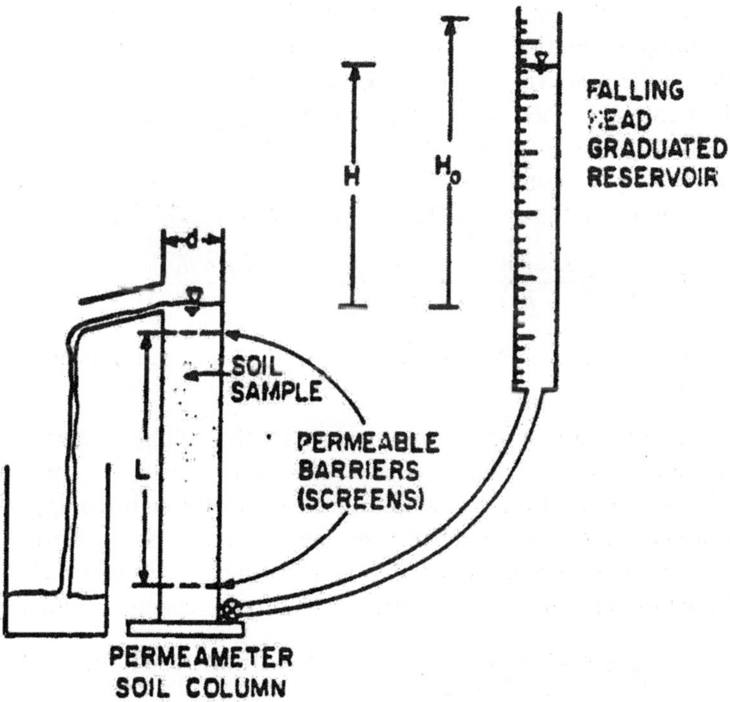

Figure 7.2 Falling head permeameter.

Darcy's equation, when rearranged to determine the hydraulic conductivity, is :

$$K = \frac{q\,\Delta l}{\Delta h} \tag{7-1a}$$

Since velocity, q, is equal to flow rate, Q (cm^3/sec), divided by flow area, A (cm^2):

$$K = \frac{Q\Delta l}{A\Delta h}. \tag{7-2}$$

In the constant head permeameter, Δl is the length of the sample, L (cm), over which the head loss, $-\Delta h$, is being measured; A (cm^2) is the column cross-sectional area; and flow rate is measured as a volume, V (cm^3), collected at the outflow over a given elapsed time, t (sec) :

$$K = \frac{VL}{tA\Delta h}. \tag{7.3}$$

The total head loss in the constant head permeameter would be, according to Figure 7-1, H_1–H_2. However, because the head loss through the tube, valves, and permeable barriers might be significant when coarse-grained soil samples are measured, piezometer taps are installed in the sides of the column so that head losses within the sample can be monitored. Using these piezometer readings and the terms shown in Figure 7-1, Equation 7-3 becomes :

$$K = \frac{VL_1}{tA\left(h_b - h_c\right)} = \frac{VL_2}{tA\left(h_c - h_d\right)} = \frac{V\left(L_1 + L_2\right)}{tA\left(h_b - h_d\right)} \tag{7.4}$$

The head loss through the inflow tube and valve is given by (H_1–H_a).

In the falling head permeameter, the supply reservoir is volumetrically calibrated, so that the water level, H, not only indicates the head on the sample (head losses through the apparatus fixtures being ignored), but also the volume passing through it between two measurements. Flow rate, which will vary with the change in depth in the reservoir, is equal to the change of volume in the reservoir with time, $-\frac{dV}{dt}$, which is equal, assuming a constant reservoir cross-sectiona area, a, to $-a\frac{dh}{dt}$. Inserting this relationship into Equation 7-2, gives :

$$-a\frac{dh}{dt} = KA\frac{h}{L} \tag{7-5}$$

Separating the variables and integrating from an initial head, H_O, to a head, H, which occurs at time, T :

$$-\int_{H_o}^{H}\left(\frac{1}{h}\right)dh = \int_{O}^{T}\frac{KA}{aL}\,dt,$$

$$-\ln\frac{H}{H_o} - \frac{KAT}{aL},$$

$$K = \frac{aL}{AT}\ln\frac{H_o}{H}. \tag{7.6}$$

The laboratory methods described can accurately determine the permeability of the sample tested. The only major experimental problems involve the boundaries-head losses through the permeable end barriers and edge effects along the column walls.

However, applying the results of these tests to field conditions usually cannot be done with such confidence because of the unrepresentativeness of the sample tested. The first problem is to collect an undisturbed sample. Because this is difficult, disturbed samples are often used. Disturbed samples will give an accurate indication of field permeabilities only for unstructured coarse soils (sands and gravels) or when the application is to disturbed soils, such as road fills or earthen dams. Even when undisturbed soil cores are carefully collected, it is still difficult to install them in a permeameter column without allowing flow paths along the column walls.

Another problem with predicting field subsurface flow with permeameter tests is that the permeability of a soil is usually nonhomogeneous (it varies from place to place), and anisotropic (it varies from one direction to another). These nonuniformities are caused primarily by the alluvial or eolian nature of most unconsolidated deposits. In cultivated fields, plow pans are an example of anisotropy. It is not uncommon for the horizontal permeability of some subsurface deposits to be 1,000 times greater than the vertical permeability.

A third problem with laboratory permeability tests is the possible non-representativeness of the fluid. Hydraulic conductivity is affected by the viscosity and density of the fluid, and these properties of water change with ion concentration and temperature. The effect of temperature on viscosity, as given in Table 7-1, is the largest and most easily predicted.

Table 7.1. Viscosity of pure water at various temperatures.

Temperature (°C)	Absolute Viscosity, η (centipoise)
0	1.79
5	1.51
10	1.31
15	1.14
20	1.00
25	0.89
30	0.80
35	0.71

Hydraulic conductivity is inversely related to viscosity. Thus, K values can be adjusted for temperature by the formula :

$$K_{st} = K_t \left(\eta_t \;/\; \eta_{st}\right) \tag{7.7}$$

where K_{st} and η_{st} are the conductivity and viscosity at some standard temperature while K_t and η_t are these properties at the measurement temperature.

Fluid properties can be eliminated from permeability measur-ements by converting the hydraulic conductivity to perme-ability, sometimes called intrinsic permeability, k:

$$k = \frac{K\eta}{\rho g}, \tag{7.8}$$

where

η = fluid viscosity in poises (dyne-sec/cm^2),

ρ = density (g/cm^3), and

g = acceleration of gravity (dyne/g).

Permeability has the dimensions of cm^2, although square microns, μ^2, $\left(\mu^2 = 10^{-8} \times cm^2\right)$ is a more common unit due to the small values.

Because of these difficulties, several field methods of determining in-situ permeabilities and subsurface flow characteristics are used which are less precise but often more representative. These methods include :

1. Infiltration tests to measure the rate of water intake at the soil surface.
2. The auger hole method, to determine the permeability of shallow saturated strata; and
3. Well tests to determine permeabilities of water bearing aquifers.

The first two methods are described in the following two exercises.

7.2 EQUIPMENT (see Figs. 7-3 and 7-4)

1. Constant head reservoir (can be an overflow reservoir as shown in Figure 7.1, or a Marriotte syphon (see Exercise 9)
2. Falling head graduated reservoir
3. Soil column with bottom inflow tap, top overflow, and piezomete; taps
4. Screens or other permeable barriers
5. Piezometer board
6. Graduated cylinder
7. Column packing device
8. Funnel

Also : stop watch and meter stick

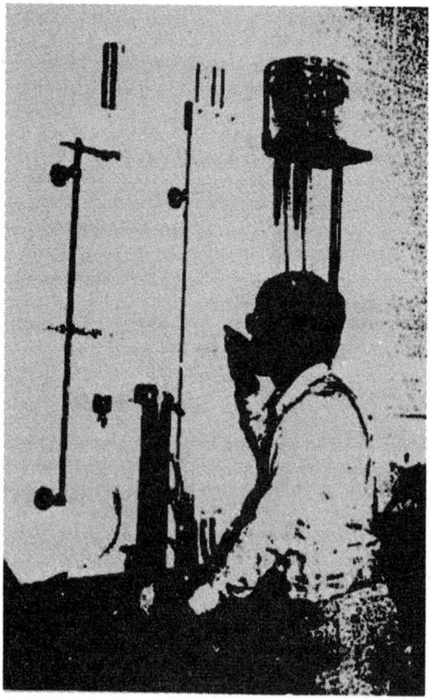

Figure 7-3. Permeameter showing (left to right) falling-head reservoir, soil column, piezometer board, and constant-head reservoir.

Figure 7-4. Permeameter soil column showing bottom valved inflow, top overflow, and piezometer taps.

7.3 PROCEDURE

Preparing the Soil Column

Disturbed sand will be used in the soil column because the high permeability of sand will speed up and make easier the measurement process and undisturbed samples are difficult to collect and prepare. Sand grainds, when poured into the column, will segregate and the packing of the grains will be uneven, causing the flow of water through the column to also be uneven. Consequently, a ``packing'' device shown in Figure 7.5, is used to allow more uniform and consistent packing of the columns. The screens in the packing device, which should be held only a few centimeters. Screens should be used in the column above and below the sample to hold the sample in place while allowing water to pass freely.

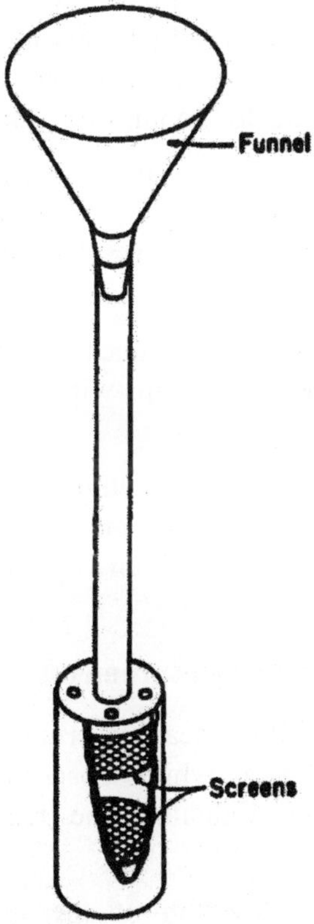

Figure 7.5. Permeability column packing device.

1. Place a screen on the bottom permeameter column barrier.
2. Fill the column to about 4 cm below the overflow outlet using the packing device. Keep the bottom of the packing device no more than 6 cm above the level of the sand in the column.
3. Place a screen on top of the sample and clamp the top retaining barrier in place.

Constant Head Permeameter Measurements

1. Attach the hoses from the reservoir and the piezometers to the soil column. Close the column inflow valve.
2. Fill the reservoir and regulate the inflow so a small amount of water is overflowing to the drain.
3. Partially open the column inflow valve and allow the sample to be slowly wetted. Observed the evenness of the wetting front as it moves up through the sample. Unevenness indicates nonuniform packing.
4. Measure the elevation of the reservoir water level, H_1, the column overflow level, H_2, and the piezometer scales relative to a datum. Measure the column diameter, d, and sample lengths, L, L_1 and L_2.
5. When the sample is wet, check that the piezometers are all working. Open plugged taps with a fine wire and eliminate air bubbles in the lines.
6. When the permeameter column is overflowing steadily, take piezometer readings and collect the overflow in a graduated cylinder for a known period of time. Repeat the measurement.

Falling Head Permeameter Measurements

1. Calibrate the falling head reservoir by collecting known volumes of water while observing readings off the calibrated scale. Calculate the reservoir cross-sectional area, a (cm^2).
2. Determine the elevation of the zero reservoir reading relative to the column outflow. Measure the sample length, L, and cross-sectional diameter, d.

3. If the sample is dry, saturate it as explained in #3 under constant-head measurements.

4. Open the column inflow valve completely and record reservoir readings over time as the reservoir level is falling through its whole range. Refill the reservoir and repeat the measurement.

Calculations

1. Plot hydraulic gradient, $-\Delta h / \Delta l$, vs. flow rate, q, for the constant-head measurements to determine whether they are proportional at all the measured flow rates. They are proportional if related by a straight line with a zero intercept. The slope of the line is the hydraulic conductivity.

2. Calculate the hydraulic conductivity using the falling head data with Equation 7-6 for the whole fall and several intermediate readings and compare the values. The hydraulic conductivity should also be determined by fitting a straight line through time and head loss data plotted on semi-log paper (with H on the log scale). The slope of the line, m, is equal to $-\dfrac{aL}{AK}$, so :

$$K = -\frac{aL}{Am} \tag{7.9}$$

Slope, m, can be found most easily by dividiing the time interval over which H decreases by one log cycle by –2.3 (–In 10).

7.4 ANALYSIS AND DISCUSSION

Table 7-2 lists some permeability classes for soils. In which classes would the tested sands fall?

Table 7-2. Permeability classes of saturated soils for ranges of hydraulic conducitivities (from Table 13-1 in Black *et al.*, (1965)).

Class	Hydraulic Conductivity to Water K (cm/sec)
Very slow	<0.002

Slow	0.002 – 0.008
Moderately slow	0.008 – 0.033
Moderate	0.033 – 0.100
Moderately rapid	0.100 – 0.210
Rapid	0.210 – 0.420
Very rapid	> 0.420

Assuming groundwater is at about 10°C and the tap water in the lab is at 15°C, what would be the hydraulic conductivity of the measured soils to groundwater?

Convert the measured hydraulic conductivities to intrinsic permeabilities, k, using Eq. 7-8 and Table 7-1 and 7-2. Assume a water temperature of 15°C.

If the measured conductivities varied with flow rate or measurement method, discuss the possible reasons.

7.5 REFERENCES

American Society for Testing and Materials (ASTM). 1980. Permeability of Granular Soils (Constant Head) (ASTM D2434-68), in Annual Book of ASTM Standards. ASTM, 1916 Race St., Philadelphia, PA.

Black, C.A., D.D. Evans, J.L. White, L.E. Ensminger, and F.E. Clark. 1965. Methods of Soil Analysis, Part 1, Chap. 13, Laboratory Measurement of Hydraulic Conductivity of Saturated Soil, by A. Klutte, ASA Monograph Series No. 9. American Society of Agronomy, Madison, WI.

Bureau of Reclamation (USBR). 1974. Earth Manual. Appendix E-13 : Permeability and Settlement of Soils. U.S. Dept. of Interior, USBR, Second Ed.

Commonwealth Buraue of Soils. 1974. Methods of Analysis of Irrigated Soils, J. Loveday ed. Chap. 10, Hydraulic Conductivity, by D.S. McIntyre and J. Loveday. Technical Communication No. 54. Commonwealth Agricultural Bureau, Farnham Royal, Bucks, England.

Taherian, M.O., J.W. Hummel, and E.C. Rebuck, 1976. Comparison of Different Techniques of Measuring Soil Hydraulic Conductivity. In Proceedings, Third National Drainage Symposium, Chicago, IL. Dec. 1976. American Society of Agricultural Engineers, St. Joseph, MO.

Soil-Water Engineering Laboratory
Exercise 7
Constant Head Permeameter Data Sheet

Name ____________________

Date ____________________

Soil Sample Description ____________________

Elevation of constant head reservoir above datum, H_1 (cm) ____________________

Elevation of permeameter column overflow above datum, H_2 (cm) ____________________

Elevation of piezometer board reference (0 reading) above datum, h_o (cm) ____________________

Permeameter column diameter, d (cm) __________ ; column area, A (cm^2) __________

Length of soil sample in column, L (cm) ____________________

Length of soil sample between piezometer taps b and c, L_1 (cm) ____________________

Length of soil sample between piezometer taps c and d, L_2 (cm) ____________________

Volume Collected V (cm^3)	Elapsed Time t (sec)	Piezometer Readings				Flow Rate Q (cm^3/sec)	Volume Flux q (cm/sec)	Hydraulic Gradient $-\Delta h/L$			Hydraulic Conductivity K (cm/sec)
		h_a (cm)	h_b (cm)	h_c (cm)	h_d (cm)			$\frac{h_b-h_d}{L_1+L_2}$	$\frac{h_b-h_c}{L_1}$	$\frac{h_c-h_d}{L_2}$	

Soil-Water Engineering Laboratory
Exercise 7
Falling Head Permeameter Data Sheet

Name________________________

Date________________________

Soil Sample Description________________________

Elevation of reservoir reference (0 reading) above datum (cm)__________

Elevation of permeameter column outflow above datum (cm)______________

Graduated reservoir reference elevation relative to the column overflow (cm)____________________

Graduated reservoir cross-sectional area, a, (cm^2)____________________

Permeameter column diameter, d, (cm)______________________________

Column area, A (cm^2)________________________

Sample length, L (cm)____

Reservoir Reading (cm)	Elapsed Time T (sec)	Reservoir Elevation H (cm)	$\ell n \frac{H_o}{H}$	Hydraulic Conductivity K (cm/sec)

EXERCISE 8

FIELD MEASUREMENT OF HYDRAULIC CONDUCTIVITY BY THE AUGER-HOLE METHOD

8.1 INTRODUCTION

Due to the difficulty of collecting and preparing representative soil samples for hydraulic conductivity measurements in laboratory permeameters, a field method is desirable. The auger hole method is a quick and simple field technique with minimal equipment requirements for determining the hydraulic conductivity of porous strata below a water table. The method basically entails pumping or bailing out a hole which extends belwo the water table and measuring how quickly the hole refills. Although the flow process is too complicated to allow solution by the type of simple equations which are used with one-dimesnional steady flow, the process has been studied and modeled and coefficients have been developed which relate the soil hydraulic conductivity, K to the rate at which the hole fills.

The method is limited by the fact that a water table must exist at the level of the soil layer which is to be measured. This sometimes requires waiting until specific times when the water table is high, such as in the spring or part way through an irrigation season. Methods have been developed for measuring K above a water table (Black *et al.,* 1965, Section 15), but they are more difficult and required more equipment. The primary use of auger hole measurements is to determine hydraulic conductivities for drainage design, in which case high water tables usually exist.

Soils consist of layers which have varying hydraulic conductivities. The auger hole measurements tend to be biased toward the horizontal conductivity of the more permeable layers. By varying hole depths, permeabilities of different layers can be estimated. Vertical conductivities can be better estimated by the piezometer method in which the hole is cased and water enters only from the bottom (Black *et al.,* 1965, Section 14).

Some advantages of using the auger hole method over laboratory methods include :

1. The soil is not disturbed (original conditions in the field),
2. The ``sample" is large and takes into account many types of water channels in the soil, and
3. The fluid used is the soil solution itself and not tap water or distilled water having unknown effects.

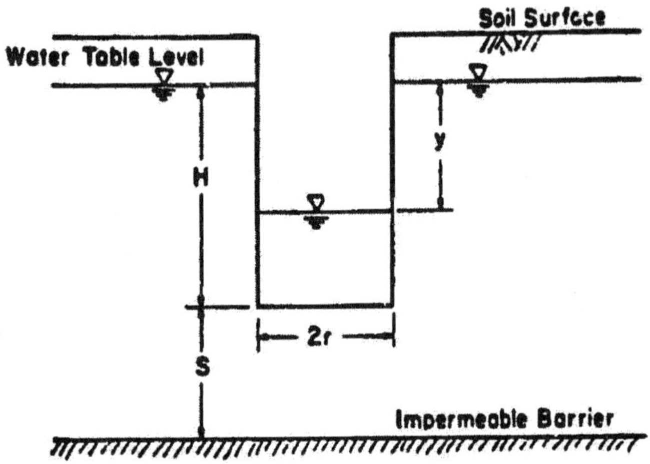

Figure 8-1. Required dimensional parameters for auger hole tests.

Figure 8-1 shows the required geometric parameters for auger hole measurements. In the figure, r is the radius of the hole, H is the depth of the hole below the water table, or equivalently the depth of water in the hole before it is pumped, y is the level of the water surface in the hole below the water table after it has been pumped, and S is the distance from the bottom of the hole to an impermeable barrier (such as a clay layer or bedrock). Impermeable layers can be found by augering the hole deeper after the measurement is complete and observing the texture and density of the removed soil. Fortunately, the method is relatively insensitive to impermeable barriers located farther than H below the bottom of the hole.

The relationship between the observed rate of rise of water in the auger hole and the hydraulic conductivity is expressed by :

$$K = \frac{C}{864}\frac{dy}{dt}, \tag{8-1}$$

Where C is the shape factor [C = f (H/r, y/H, S/H)] of the auger hole, dy is the change in water height in the auger hole (cm) that occurs in time dt (sec), and K is the hydraulic conductivity (cm/sec). The constant 864 in a unit conversion factor.

The values of C given in Table 8-1 were obtained using auger hole seepage theory (Boast and Kirkham, 1971).

Table 8-1. Values of C for Equation 8-1 for auger holes underlain by impermeable material at given depths (Boast and Kirkham, 1971).

M/r	y/H	S/H								
		0	0.05	0.1	0.2	0.5	1	2	5	00
1	1	447	423	404	375	323	286	264	255	254
	0.75	469	450	434	408	360	324	303	292	291
	0.5	555	537	522	497	449	411	386	380	379
2	1	186	176	167	154	134	123	118	116	115
	0.75	196	187	180	168	149	138	133	131	131
	0.5	234	225	218	207	188	175	169	167	167
5	1	51.9	48.6	46.2	42.8	38.7	36.9	36.1		35.8
	0.75	54.8	52.0	49.9	46.8	42.8	41.0	40.2		40.0
	0.5	66.1	63.4	16.3	58.1	53.9	51.9	51.0		50.7
10	1	18.1	16.9	16.1	15.2	14.1	13.6	13.4		13.4
	0.75	19.1	18.1	17.4	16.5	15.5	15.0	14.8		14.8
	0.5	23.3	22.3	21.5	20.6	19.5	19.0	18.8		18.7
20	1	5.91	5.53	5.30	5.06	4.81	4.70	4.66		4.64
	0.75	6.27	5.94	5.73	5.50	5.25	5.15	5.10		5.08
	0.5	7.67	7.34	7.12	6.88	6.60	6.48	6.43		6.41
50	1	1.25	1.18	1.14	1.11	1.07	1.05			1.04
	0.75	1.33	1.27	1.23	1.20	1.16	1.14			1.13
	0.5	1.64	1.57	1.54	1.50	1.46	1.44			1.43
100	1	0.35	0.35	0.34	0.34	0.33	0.32			0.32
	0.75	0.38	0.38	0.37	0.36	0.35	0.35			0.35
	0.5	0.47	0.47	0.46	0.45	0.44	0.44			0.44

When the actual measured parameters are not represented on the table, they must be interpolated between given values. Graphical interpolation on log-log paper is most precise.

8.2 EQUIPMENT

1. Soil auger or a drilling rig, such as r ``Giddings" rig.
2. Permeable material for casing the auger hole to prevent caving such as thin perforated sheet metal pipe, perforated stove-pipe, drain tile, or thin perforated PVC pipe, plus fine gravel to pack the hole. If the soil is sufficiently cohesive and the hole is to be used only once, casing is not required.
3. Water elevation indicator. The indicator device used in this exercise, shown in Figures 8-2, 8-3, and 8-4, consists of a float lowered into the auger hole and connected to a counter-weigh with a tape. This tape which is suspended across a table is marked at equal time intervals as the water level rises in the auger hole. Electrical resistance well loggers could also be used.
4. Pump or bailer for quickly removing water from the auger hole.

Also : stop watch

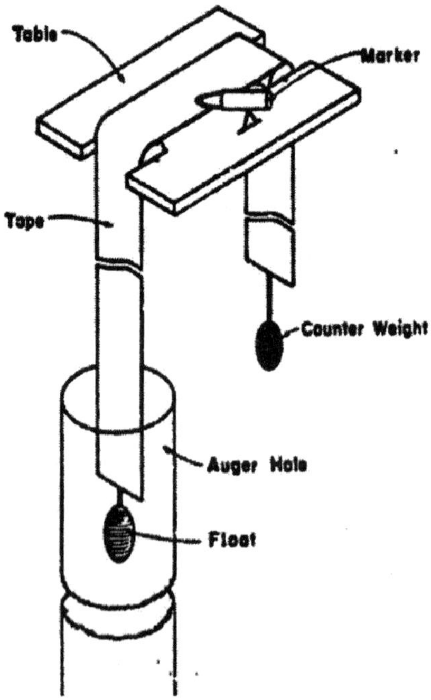

Figure 8-2. Water elevation indicator device.

Figure 8-3. Water level recorder and cased will used in the auger hole method.

Figure 8-4. Water level recorder showing table, tape, and amrking pen.

8.3 PROCEDURE

In the following discussion, the installation of the auger hole (Steps 1 through 4) is done before measuring the hydraulic conductivity. Successive steps (5 throguh 9) are done several times so that more than one estimate of the hydraulic conducitivity can be made.

Installation

1. Dig a vertical hole with a soil augar to to the desired depth below the water table (at least 30 cm).
2. Install the permeable casing in the hole leaving the tope end of the casing several centimeters above the soil surface. To install the casing, place a small quantity of gravel on the bottom of the hole and set the perforated open pipe on this gravel. Back-fill around the pipe with gravel to a level above the water table. The original soil can be used to fill the remaining portion of the hole around the pipe to the soil surface. The casing should be capped for protection.
3. Pump the hole several times until clean water is obtained to "develop" it.
4. Allow the water to seep into and fill the auger hole to the level of the water table. This will require at least a day for heavy, tight soils, but only a few minutes are needed for permeably sandy soils.

Measurement

5. Measure the required geometric parameters (Fig. 8-1) and record on the data collection sheet. It should be noted that the hole diameter (2r) is the diameter of the packed bore and not the diameter of the casing.
6. When the water in the auger hole is at the water table level, lower the float into the auger hole and mark a datum point on the tape.
7. Pull the float out of the auger hole.
8. Pump or bail out the water quickly. Begin counting time when the pump is stopped.
9. Quickly lower the float into the auger hole and mark the tape at frequent and equal time intervals as the water level rises in the auger hole.

Calculations

Note: Sample data is analyzed in Figures 8-5, 8-6 and 8-7.

Soil-Water Engineering Laboratory
Exercise 8
Auger Hole Data Sheet

Name____________________

Date____________________

Location____________________

Description of subsoil strata
(note during augering of hole)____________________

Geometric Parameters

H (cm)__________

r (cm)__________

S (cm)__________

H/r __________

S/H __________

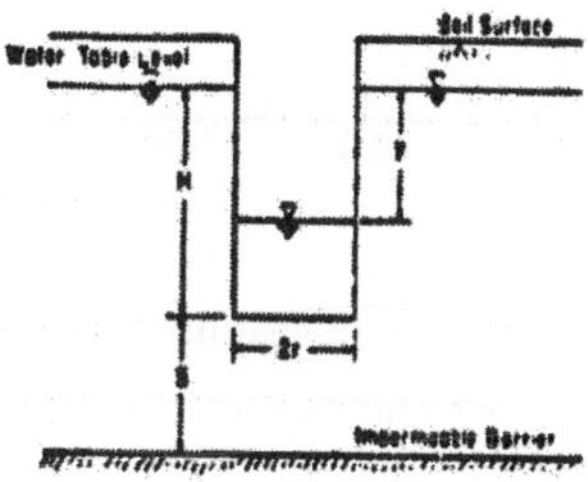

Run 1		Run 2		Run 3		Run 4	
t (sec)	y (cm)	t (sec)	y (cm)	t (sec)	y (cm)	t (sec)	y (cm)

	Run 1	Run 2	Run 3	Run 4
y				
$\frac{dy}{dt}$ (cm/sec)				
$\frac{y}{H}$				
C				
K (cm/sec)				

Figure 8-5. Sample auger hole data

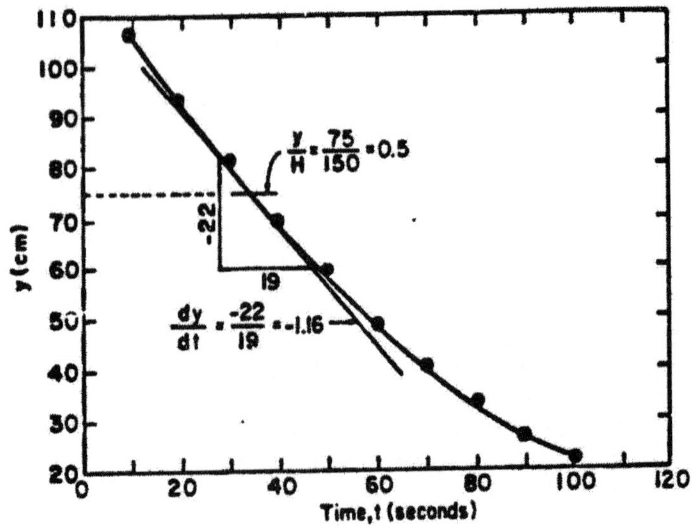

Figure 8-6. Graphical determination of dy/dt.

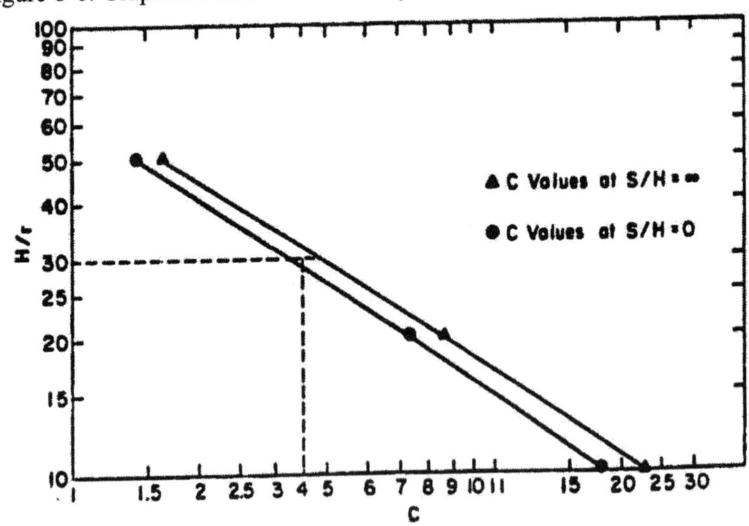

Figure 8-7. Logarithmic interpolation of C for y/H = 0.5 and H/r = 30.

1. Measure the distance from the marked water level at each time increment to the water table level on the tape and record on the data collection sheet.

2. Plot y vs. t. A a chosen y value (preferably equal to 0.75 or 0.5 H so that C values can be more easily determined from

Table 8-1) determine the slope of the curve, $\frac{dy}{dt}$. An example of this determination is shown in Figure 8-6.

3. Determine the shape factor, C, from the geomeric parameters and Table 8-1. If S is not known, estimate C for the extreme cases of S = 0 and S = ∞ to determine the range. If one of the paramter ratios is not given in the table, interpolate graphically on log-log paper. Figure 8-7 demonstrates the determination of interpolated C values.

4. Insert the C and $\frac{dy}{dt}$ values into Equation 8-1 to determine the hydraulic conductivity, K. Hydraulic conductivities are sometimes expressed in m/day which is equivalent to K (cm/sec) x 860.

8.4. ANALYSIS AND DISCUSSION

Compare the derived K values calculated at different y values and for the different runs. Determine the 95 percent confidence interval for the mean. What is the effect of the impermeable barrier on the measured hydraulic conductivity? According to the conductivity ranges given in Table 7-2 of Exercise 7, would the soil be considered highly permeable? What might the soil texture be? If the hydraulic conductivity is used for drainage design, would the temperature effects discussed in Exercise 7 need to be considered.

8.5. REFERENCES

Black, C.A., D.D. Evans, J.L. White, L.E. Ensminger, and F.E. Clark. 1965. Methods of Soil Analysis, Part 1, Chapt. 14, Field Measurement of Hydraulic Conductivity Below a Water Table, by L. Boersma. ASA Monograph Series No. 9. American Society of Agronomy. Madison, WI.

Boast, C.W. and D. Kirkham. 1971. Auger Seepage Hole Theory. Soil Science Society of American Proceedings 35-3, May June.

EXERCISE 9

INFILTRATION

9.1. INTRODUCTION

Infiltration is the entry of water into soil through the soil surface. Nearly all water which enters the soil enters via infiltration, either from rainfall or from water directed to the soil surface via irrigation. Knowledge of the infiltration characteristics of a soil is basic information required for designing efficienct irrigation system of predicting runoff. The soil infiltration rate will affect advance and recession times, deep percolation and tailwater runoff in furrow and border systems, advance and ponded times in level borders (basins), and maximum allowable application rate for sprinkler or drip irrigation systems.

The time rate at which water is absorbed by the soil is the infiltration rate, I, which has units of volume per unit area per unit time, $cm^3/cm^2/min$, or equivalent depth per unit time, cm/min. The cumulative infiltration, z, is the total accumulated depth infiltrated in a given time period. Cumulative infiltration is thus the integration of the infiltration rate, and conversely, I is the derivative of z.

$$z(t) = \int_{O}^{t} I \, dt \tag{9.1}$$

$$I(t) = \frac{dz}{dt} \tag{9.2}$$

The rate at which water infiltrates into soils usually decreases over time until a relatively steady basic infiltration rate is asymptotically approached. The decrease is caused by the decreasing capillary pressure of the soil as it becomes wet. The constant rate is due to the constant gravity force, the other component of the driving force.

Figure 9.1 shows typical infiltration rate and cumulative infiltration curves for an agricultural soil.

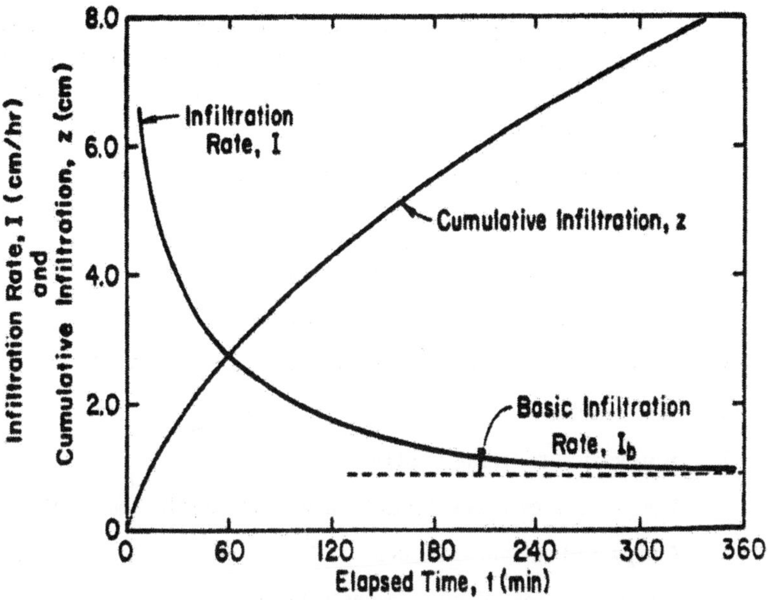

Figure 9.1. Typical infiltration rate and cumulative infiltration curves.

Due to the changing driving forces (hydraulic gradient) and the varying moisture content of the soil, the infiltration rate of a soil cannot be directly related to its hydraulic conductivity. This is why a specialized measurement must be made. The infiltration rate of soil will, however, vary with the saturated hydraulic conductivity of the soil layers. Layers with low conductivities, either at or below the surface, will limit the infiltration rate especially the basic rate. Consequently, surface crusts, silt deposits, implement tracks, plow layers, or clay layers will have a major effect on infiltration.

Pre-existing soil moisture content will affect the initial infiltration rate. Drier soils will have higher initial rates, due to their high capillary pressure. Saturated soils initially infiltrate water at near the basic rate.

Infiltration rates within one area or field are highly variable-variations as high as an order of magnitude in one field are non uncommon due to inhomogeneities in the soil (both textural and structural) and variations in the above-mentioned factors. This variation is both areal and temporal. The infiltration rate of a wheel-track furrow is usually much less than that of a less compacted furrow. The intake rate during the first irrigation after cultivation is usually higher than during later irrigations. Consequently, times and locations for infiltration measurements must be chosen carefully in order for the results to be representative and meaningful. Some factors to consider include:

1. Soil moisture content at time of irrigation
2. Soil texture
3. Subsurface strata (plow pan, clay layers)
4. Surface soil compaction (wheel tracks)
5. Surface sealing (sedimentation, erosion, dispersion)
6. Soil cracking
7. Soil structure (tillage operations)
8. Irrigation method (sprinkler vs. surface; border vs. furrow)
9. Crops and surface mulches
10. Soil and water salt ion concentrations
11. Soil and water temperature

Due to the variability, several measurements are required to achieve a reliable estimate of the infiltration rate for a field.

Several machematical realationships or models have been proposed for infiltration rate and cumulative infiltration, including the empirical Kostiakov-Lewis equation:

$$z = Kt^{a}, \tag{9.3}$$

the theoretically based approximation developed by Phillips :

$$z = St^{1/2} + At, \tag{9.4}$$

the empirical relationship used by the U.S. Soil Conservation Service:

$$z = Kt^{a} + c, \tag{9.5}$$

and the more general empirical equation (sometimes called the modified Kostiakov-Lewis) :

$$z = Kt^{a} + I_{b}\,t, \tag{9.6}$$

where :

K and a = empirically determined coefficients,

I_b = basic infiltration rate,

S and A = theoretically determined values,

c = 0.6985 cm, and

t = time excess water is available at the surface (intake opportunity time).

In this exercise, Equation 9.6 will be used.

Several methods can be used to determine the infiltration characteristics of a soil under field conditions. Methods frequently used for basin and border irrigation are:

1. Cylinder (ring) infiltration, and
2. Basin infiltration.

The two common methods used for furrow irrigation are:

3. Blocked furrow infiltrometer, and
4. Inflow-outflow measurements on a segment of a furrow.

An additional method applicable to both border and furrow irrigation is :

5. Volume balance techniques based upon the rate of advance of the water front and estimated or measured values of the volume of water in surface storage.

Cylinder and blocked furrow infiltrometers will be used in this exercise due to their ease of use, precision, common usage, and simplicity.

Cylinder Infiltrometer

The cylinder or ring infiltrometer (Figs. 9.2 and 9.3) is a metal cylinder which is driven into the soil. It measures primarily the vertical rate of water movement into the soil surface (one-dimensional) from the pond it encloses. Cylinder infiltrometer infiltration data are obtained by measuring the depth of ponded water inside the cylinder at time intervals.

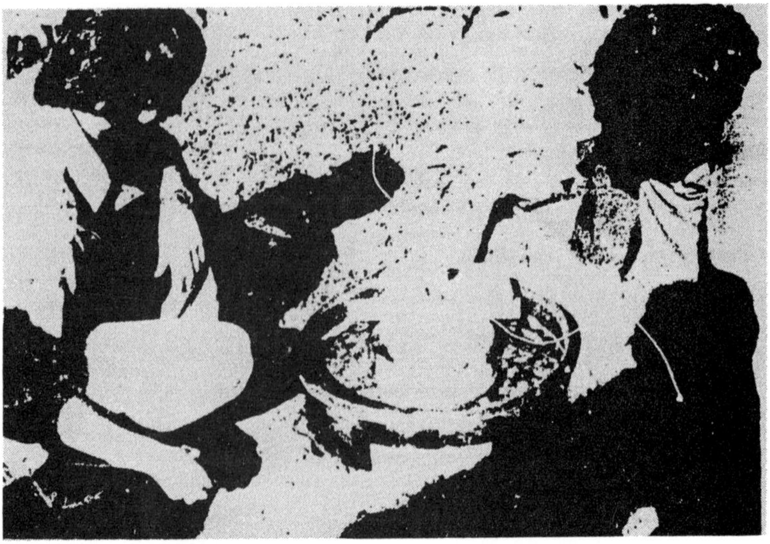

Figure 9.2 Cylinder infiltrometer in use.

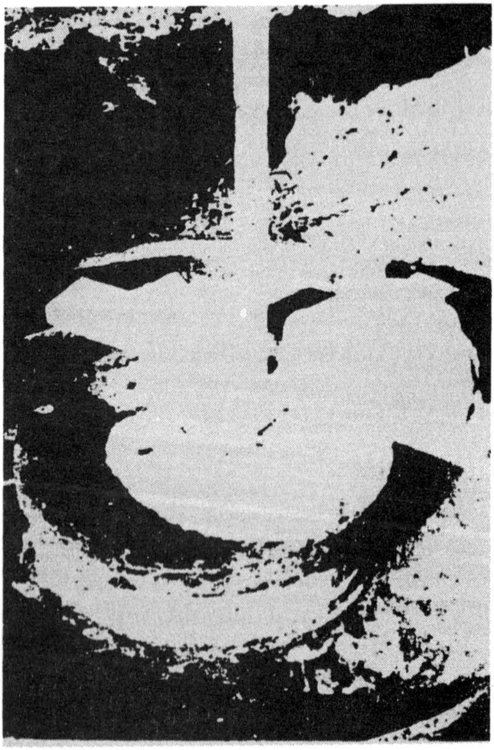

Figure 9.3 Cylinder infiltrometer showing buffer pond and depth gauge.

After water penetrates the soil to the depth of the bottom of the cylinder, it will begin to spread radially as well as vertically and the infiltration rate will change accordingly. Buffer ponds surrounding the cylinder are used to minimize this effect. Buffer ponds can be constructed by forming an earthen dike around the cylinder or by driving a larger diameter cylinder into the soil concentric with the cylinder infiltrometer. Water is maintained in the area between the two cylinders at about the same depth as that in the inner cylinder.

Two problems may make infiltration rates measured by cylinder infiltrometers unrepresentative of true infiltration rates. One is that the soil may be disturbed when the rings are driven into the ground. The second is that the ponded water may not duplicate real conditions. For example, although infiltration rates are not too sensitive to the water pressure at the surface, or equivalently the depth of water on the surface, water pressure will have some effect. The infiltration rate measured with an infiltrometer with 20 cm of ponded water will not be equivalent to that in a border with 5 cm water depth, or to that under a sprinkler where no water is ponded. Also, air which is trapped under a level basin may escape laterally from around the infiltrometer. Finally, under real conditions, water usually moves across the soil surface rearranging surface particles. This erosion and aggregation which does not occur in ponded infiltration tests tends to seal soil surfaces and reduce infiltration rates.

Blocked Furrow Infiltrometer

One of the commonly used methods for measuring the infiltration rate in furrow irrigation is the blocked furrow infiltrometer, shown in Figure 9.4. The blocked furrow infiltrometer measures the infiltration of water into the soil profile from a short segment of a furrow (about 1 meter). The water is ponded in the section by barriers driven into the soil across the furrow. Furrow infiltration is two-dimensional because it includes both vertical and horizontal movement of water into the soil. After the water penetrates the soil surface it will begin to spread laterally as well as vertically. In order to prevent excess lateral movement of water and simulate the actual irrigation, furrows adjacent to the one tested are filled with water. These are known as buffer furrows.

Figure 9.4. Blocked furrow infiltrometer with a Marriott siphon water supply reservoir. Notice the buffer furrows.

Since the wetted surface will vary as the water depth in the furrow varies, maintaining the water level in the furrow near the normal depth is important. The normal flow depth should be determined, if possible, while the furrow is being used during a normal irrigation. The constant depth can be maintained easily with a device called a Marriott syphon, shown in Figure 9.5. A Marriott syphon is a closed container with the only access to atmospheric pressure being through an adjustable tube which extends below the water surface. The air tube creates a free water surface at atmospheric pressure at the level of its bottom regardles of the actual depth of water in the container. By placing the water outlet hose in the bottom of the blocked furrow and positioning the lower end of the air tube at the desired furrow water surface elevation, the outflow from the reservoir will maintain the desired furrow water depth while allowing the volume of water added, and thus the volume infiltrated (since the volume stored in the furrow is constant) to be measured.

The main problem with blocked furrow infiltrometer measurements is that they are made with ponded water. Water running down furrows will erode and deposit surface sediments which can cause furrow cross section and surface infiltration characteristics to change.

Calculations

Infiltrated depth or volume over time is measured. The volume infiltrated into the blocked furrow can be converted to depth by dividing by the blocked furrow length and row spacing (or, if every second row irrigation is being practiced, by double the row spacing). The data should be plotted in the form of a cumulative infiltration curve, which is a plot of the accumulated infiltrated depth, z, vs. time from the start or the intake opportunity time, t. This plot will indicate obvious errors in the data. The infiltration rate at any time, which is the derivative of the cumulative infiltration, can be determined from the slop of the z vs. t curve at any time. Both types of curves are shown in Figure 9.1.

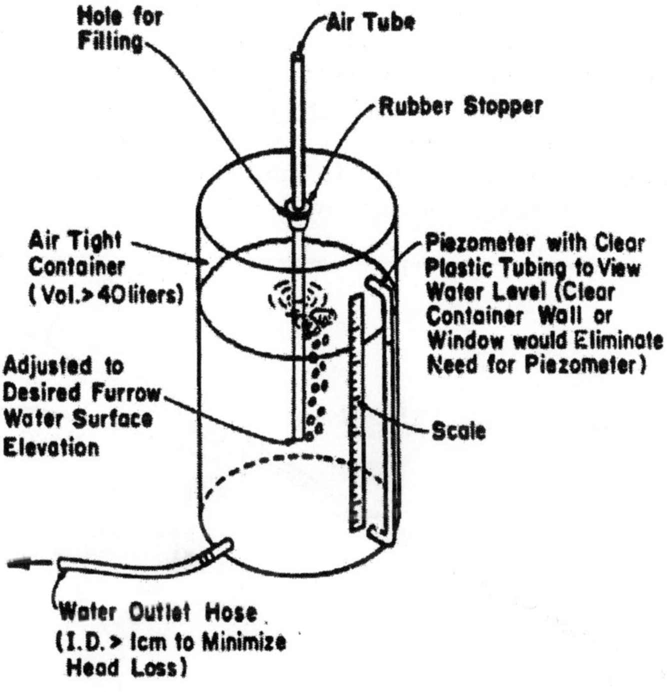

Figure 9.5. Marriott siphon water supply reservoir for blocked furrow infiltrometer.

Infiltration characteristics are easier to describe and use in the form of mathematical relationships. Therefore, the data should be fitted to Equation 9.6. This can be done graphically using log-log paper, or analytically with regression techniques.

Graphically, a power curve of the form

$$y = Kt^a \tag{9.7}$$

should plot as a straight line on log-log paper, with a as the slope of the line and K as the y value at t = 1. In Equation 9.6, y would be equal to ($z - I_b t$). Consequently, a value for I_b must be estimated before the data can be plotted. As was stated, I_b is the basic infiltration rate, so the first estimate for I_b is the infiltration rate at the end of the measurement period which can be determined from the infiltration rate plot or the slope of the cumulative infiltration vs. time curve. From this initial estimate, the value of I_b should be adjusted by trial and error until the ($z-I_b$ t) vs. t data plots as near a straight line as possible on log-log paper. A plot which curves upward, or is concave upwards, especially at longer times, indicates an I_b value which is too small, and vice versa. This graphical adjustment procedure is illustrated in Figure 9.6. If the line curves at short times, a constant probably needs to be subtracted from (if concave upwards) or added to all z values. The need for such a correction would indicate that the initial experimental conditions were faulty (i.e. plastic sheet not down tight).

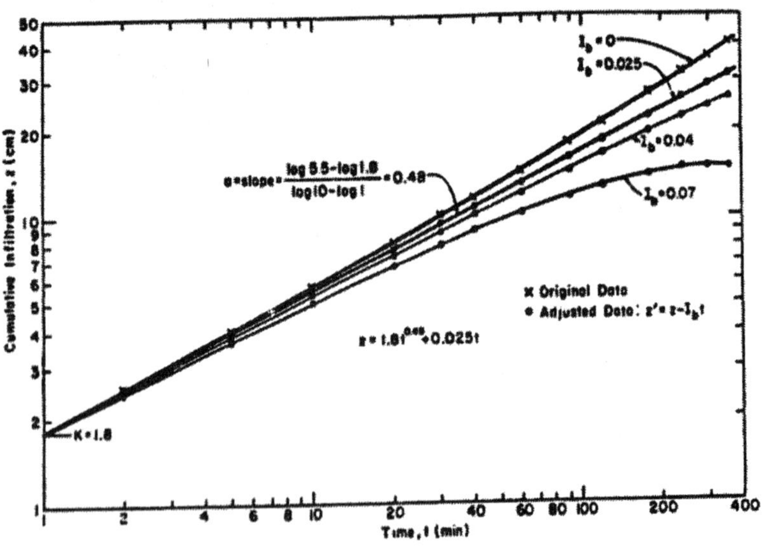

Figure 9.6. Graphical determination of general infiltration equation.

If a calculator with regression analysis capabilities is available, this curve fitting process can be done analytically. By regressing

$\ln(z - I_b\, t)$ vs. ln t, the derived slope value will be the exponent, a; and the exponential of the intercept will be coefficient K. The I_b value should be adjusted until the minimum coefficient of determination, r^2, is achieved. Regression determinations should always be checked graphically.

9.2 EQUIPMENT

1. A cylinder infiltrometer, 20 to 30 cm in diameter and 30 to 40 cm in length, made of smooth steel, rigid enough to allow it to be driven into the ground, but still thin enough to enter the soil with minimum disturbance. Two millimeter thick (14 gauge) steel will usually work, but may need reinforcement aroung the upper edge. The larger the diameter of the cylinder and the deeper the cylinder penetrates the soil, the less will be the edge effects and the greater will be the accuracy of the measurements.
2. A buffer cylinder having a diameter at least 30 cm greater than the infiltrometer and a length of about 20 cm. Construction should be similar to that of the infiltrometer. Sections of 55 gallon drums can be used as buffer cylinders. Alternately, the buffer ``ring" can also be a diked area surrounding the cylinder infiltrometer.
3. A water level gauge for measuring the changes of water level in the cylinder infiltrometer. A simple hook gauge is shown in Figure 9.7, although other arrangements are possible.
4. Two metal sheets, about 90 cm wide, 30 cm high and 0.20 cm thick, with reinforcement at the upper end.
5. Marriott syphon or other regulated supply reservoir with attached water level gauge.
6. Staff gauge or other means to indicate the constant water level required in the furrow.
7. Equipment for installing the cylinders and furrow dams such as a metal plate or heavy timber and a sledge hammer.
8. A plastic sheet or other waterproof membrane.
9. Source of water (about 200 liters for each ring infiltrometer and 800 liters for each blocked furrow test).

10. Bucket

Also : shovel, watch.

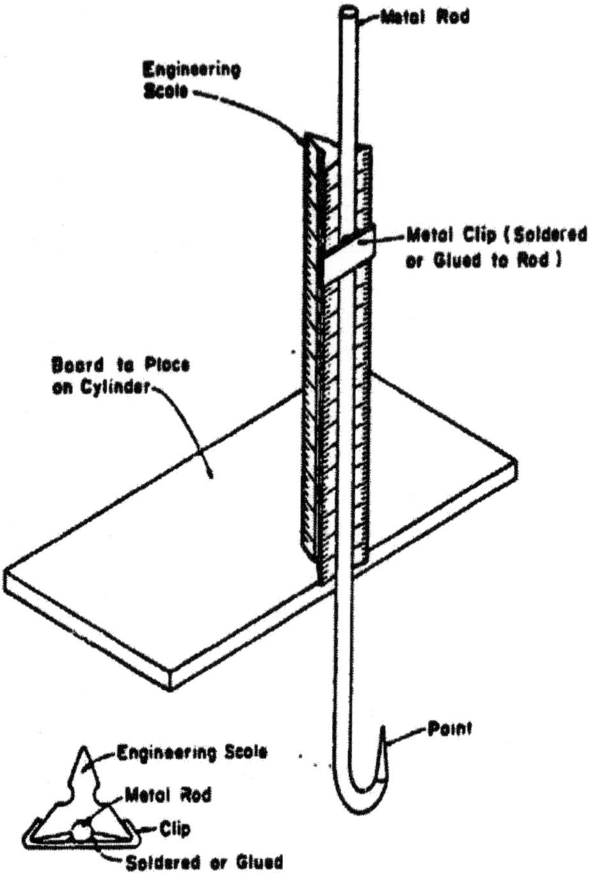

Figure 9.7. Hook gauge for cylinder infiltrometer.

9.3 PROCEDURE

Cylinder Infiltrometer

1. Select a representative location for each cylinder and examine carefully for signs of unusal surface disturbances, such as stones that might damage cylinders, cracks which might give nonrepresentative readings, etc. Avoid areas that may have been affected by unusual animal or

machinery traffic. Note the soil type and condition (tillage, vegetation, moisture).

2. Drive the infiltrometer cylinder to a depth of at least 15 cm. The cylinder should be installed as vertically as possible. This can be assured by checking the alignment frequently during the installation procedure. Do not drive the cylinder into the soil irregualrly so that first one side then the other goes down. This procedure produces poor contact between the cylinder wall and soil, and may disturb the soil core within the cylinder. If the cylinder gets out of alignment while driving, remove it and reset it in a comparable area nearby.

3. Set the buffer cylinder around the infiltrometer and drive it into the soil. This outside cylinder need not be driven as deep as the cylinder infiltrometer generally, 5 to 10 cm into the soil will be adequate.

4. Place a plastic sheet or other waterproof membrane on the soil within the cylinder infiltrometer so that it forms an inner cylinder to hold the water. The plastic should be in contact with the soil at the bottom of the infiltrometer and extend up the walls of the infiltrometer at least 15 cm.

5. Fill the buffer pond with water to a depth roughly equal to the depth desired in the inner ring. Maintain roughly equivalent depths throughout the period of observation.

6. Fill the cylinder infiltrometer with water to a depth of about 10 cm. In extremely porous (high intake) soils, a greater depth may be needed.

7. Place the hook gauge board on the cylinder and set the point of the gauge at the water level. Read the scale from the top edge of the clip. This is the initial depth reading and will correspond to the initial time reading.

8. Quickly, but gently, remove the plastic membrane recording the time at which this is done. This is the initial time reading. Make a hook gauge reading as soon as the water surface is still enough to do so, and compare this reading to the initial one to insure that there were no air spaces below the plastic.

9. Make additional hook gauge readings at periodic intervals and record the hook gauge and time readings. Intervals between observations should be short (2 to 5 min) at the start of the test. After 20 minutes the intervals may be increased to 10 to 20 min. After about 2 hours, measure-ments made at 30 to 50 minute intervals will usually be sufficient.
10. When the water level has dropped 4 to 5 cm in the cylinder infiltrometer, carefully add a sufficient volume of water to return the water surface to the approximate initial level. The known volume of water added, divided by the cross-sectional area of the cylinder, gives the depth of water added to the cylinder. If a volumetric container is not available, the added depth can be closely estimated by plotting the depth readings over time just before and after the water was added and determining the projected depth increase at the time the water was added. The volume added and corresponding depth added, or the projected depth added should be recorded on the data collection sheet.
11. Where abnormally high or low infiltration values are indicated by the test results, the infiltrometer should be dug out and the soil examined for possible causes. If, while running the test, an obvious cause for a high rate of infiltration can be seen, such as water rising outside the cylinder, the test should be terminated immediately and the cylinder moved to a nearby location and reinstalled.
12. The measurements should be continued until the change in depth over time becomes fairly constant. The time required will generally vary from 3 to 6 hours, depending on the preexisting moisture conditions and soil hydraulic conductivity.

Blocked Furrow Infiltrometer

1. Select a representative furrow section being cognizant of furrow shape, soil condition, and whether tractor or implement wheels have compacted the soil.
2. Drive two metal sheets across the furrow about 1 meter apart vertically into the soil to a depth of approximately 15 cm. Disturb of the soil surface as little as possible.

3. Similarly, install metal sheets or build earthen dams in the parallel buffer furrows. If every row irrigation is being used, the adjoining furrows would be the buffer furrows. If alternate furrows are irrigated, the second furrow on each side would be the buffer furrows.

4. Cover the bottom of the test furrow with a plastic sheet or other impermeable material.

5. Fill the test furrow with water to the level at which water would normally flow during irrigation.

6. Smooth a level platform in the bottom of the test furrow outside the ponded section. Set the Marriott syphon on the platform. Clamp off the outlet hose (or hold the end above the top of the container) and fill the container. Insert the air tube and stopper and adjust the bottom of the tube to a level a little below the elevation of the water surface in the furrow. Lay the outlet hose in the furrow and let the water flow until the air stops bubbling. This water flow is necessary to establish the vacuum in the top of the container required for the reservoir to operate properly. Then slowly raise the air tube until water just begins to flow (or air bubbles) again. The tube bottom should now be at the same elevation as the water surface in the furrow. Record the water level in the reservoir and note the level in the furrow.

7. Fill the buffer furrows with water and maintain the level at roughly the same depth as in the measured furrow throughout the test.

8. Quickly, but gently, remove the plastic sheet and begin recording time.

9. Record the reservoir water level with time. When the water level approaches the bottom of the air tube, clamp off the outlet hose, record the depth, remove the stopper (without disturbing the tube setting), refill the reservoir, reinsert the air tube and stopper, record the new depth, and open the outlet.

10. If the water level in the furrow varies above or below the desired level, adjust the air tube in the syphon accordingly. During the early portion of the test when infiltration rates

are high, it may be necessary to set the tube slightly above the furrow water level in order to create the head required to generate the required flow through the outlet hose. Temporarily removing the stopper will also increase flow rates to refill the furrow.

11. Continue the readings as explained in procedures 9 and 12 under Cylinder Infiltrometer above.
12. Determine the cross-sectional area of the reservoir and the furrow spacing. Reservoir depth readings are converted to volumes by multiplying by the reservoir cross-sectional area and converted to infiltrated depths by dividing the cumulative volume by the product of the furrow length and the furrow spacing.

Calculations

1. Calculate cumulative depths by subtracting the present level reading from the initial reading for the previous refilling, and adding it to the cumulative depth before the previous filling; or by subtracting the present level reading from the sum of the initial level reading and all equivalent depths added. Blocked furrow reservoir cumulative depth data is converted to volumes by multiplying by reservoir cross-sectional area, and to infiltrated depths by dividing the volumes by irrigated area.
2. Plot both blocked furrow and cylinder infiltrometer data (cumulative infiltrated depth vs. elapsed time) on both rectangular and log-log paper. Equation 9-6 should be fit to both data sets graphically, as demonstrated in Fig. 9.6, and if a computer or programmable calculater is available, using the power curve regression techniques listed in Appendix 2.
3. Determine infiltration rates by taking graphical derivative of the cumulative infiltrated depth curve and the mathematical derivative of the derived cumulative infiltration equation. The graphical derivative is the slope of the curve in rectangular coordinates. The differentiated equation should be plotted with the graphically derived curve to compare the methods.

9.4 ANALYSIS AND DISCUSSION

Would you expect these tests to over- or underestimate the infiltration rate which actually occurs under normal irrigation conditions? Why? Consider the facts listed in the Introduction plus the physical measurement process. Would you expect the infiltration rates to vary throughout the season as water temperatures vary? How much? (See Exercise 7).

The U.S. Soil Conservation Service (SCS) has run many cylinder infiltration tests. They have adopted the mathematical relationship :

$$z = Kt^a + c \tag{9.5}$$

where c = 0.6985 cm as best fitting their data, and have categorized soils into the intake families shown in Figure 9.8. The SCS then uses these families as desin parameters for irrigation systems. Plot a few of your collected data points on Figure 9.8 to determine whether the data fits the SCS model, and the appropriate intake family.

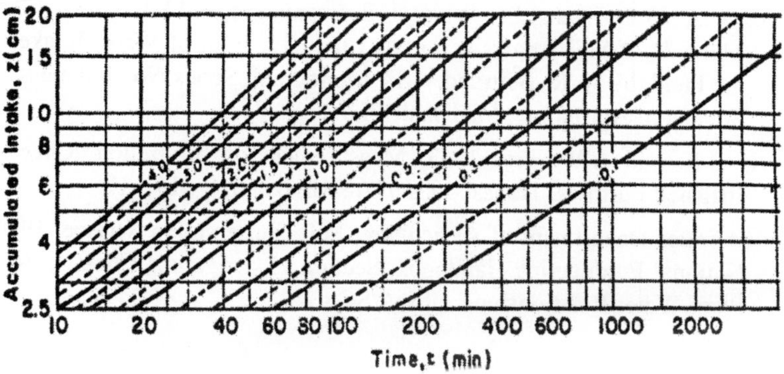

Intake Family	K	a
0.1	0.06198	0.661
0.3	0.09347	0.721
0.5	0.11862	0.756
1.0	0.17805	0.785
1.5	0.22835	0.799
2.0	0.27534	0.808
3.0	0.36500	0.816
4.0	0.44450	0.823

Figure 9.8. Soil Conservzation Service intake families (adapted from SCS National Engineering Handbook, Section 15, Chapter 1 (USDA, 1964). $z = Kt^a + c$, where c = 0.6985 cm for all intake families.

The wetted area of a furrow is less per unit area irrigated than that of a totally wetted surface as represented by a cylinder infiltrometer test. This smaller wetted surface should cause lower measured infiltration rates by the blocked furrow method. We this actually the case? If so, is the reduction as great as the reduction in wetted surface area? On the tested soil, how much longer must water run in a furrow to achieve the same irrigation as would be achieved by border irrigation?

If water is applied to this soil from a sprinkler which applies 2 cm/hr of water, how long will it take to infiltrate 8 cm depth? Will water pond on the surface? When?

9.5 REFERENCES

American Society for Testing and Materials (ASTM). 1981. Infiltration Rate of Soils in Field Using Double Ring Infiltrometers (ASTM D3385-75), in Annual Book of ASTM Standards. ASTM, 1916 Race St., Philadelphia, PA.

Black, C.A., D.D. Evans, J.L. White, L.E. Ensminger, and F.E. Clark. 1965. Mehtods of Soil Analysis, Part 1, Chap. 12, Rate of Water Intake in the Field, by A. R. Bertrand. ASA Monograph Series No. 9. American Society of Agronomy, Madison, WI.

Davis, J.R. and A.W. Fry. 1963. Measurement of Infiltration Rates in Irrigated Furrows. Transactions of ASAE, 65 : 318-319.

United States Department of Agriculture, Soil Conservation Service. 1964. National Engineering Handbook, Section 15, Irrigation, Chap. 1, Soil-Plant-Water Relationships. USDA/SCS.

Soil-Water Engineering Laboratory
Exercise 9
Cylinder Infiltrometer Data Sheet

Name____________________ **Date**____________ **Location**______________

Soil Type and Condition____

Ring Diameter (cm)____________ **Buffer Pond Diameter (cm)**____________

Time (hr/min)	Elapsed time, t (min)	Cylinder Level Reading (cm)	Volume Added (cm^3)	Equivalent Depth Added (cm)	Cumulative Depth Infiltrated z (cm)	Comments

Soil-Water Engineering Laboratory
Exercise 9
Blocked Furrow Infiltrometer Data Collection Sheet

Name__________________ Date__________ Location______________

Soil Type and Condition____

Furrow Length (cm)____________; Furrow Spacing (cm)____________

Area Irrigated (cm^2)______: Reservoir cross-sectional area (cm^2)______

Wetted Perimeter (cm)______, ______, ______, ______; Ave (cm)______

Wetted Area (cm^2)________

Time (hr/min)	Elapsed time, t (min)	Reservoir Level Reading	Added Depth (cm)	Cumulative Depth (cm)	Infiltrated Volume (cm^3)	Infiltrated Depth, z (cm)	Comments

EXERCISE 10

SOIL SALINITY

10.1 INTRODUCTION

High concentrations of ions in the soil-water solution affect plant growth by excessive ion uptake (physiological) and by effects on the soil water osmotic potential and thus water uptake by the plant. Most of the ions in soil-water solutions are derived from salts, and thus the term soil salinity is used to represent the ion concentrations. This exercise will describe methods to measure the salinity of soils. The same measurement techniques can be used to determine the salt concentration of irrigation or drainage water.

The harmful effects of salts in soils depend primarily upon the ion concentration in the soil-water solution. The concentration in turn varies with the soil moisture content. For example, the soil-water salt concentration of a soil at wilting point will be approximately double that at field capacity. Consequently, a reference soil moisture must be used to make soil salinity measurements more meaningful. Soil-water is easily extracted from soils with a high watersoil ratio, but salinity measurement at high ratios is undesirable due to salt dissolution. The extraction of soil water at low moisture contents is more difficult and time consuming. Saturation is the lowest moisture content at which soil moisture is fairly easy to extract, and is a moisture content which can be determined relatively easily. Therefore, saturation is the moisture content at which soil salinity is usually measured, and the salt concentration in an extract from a saturated soil paste is considered standard for evaluating effects on plant growth.

Ion concentrations in water can be found by determining the concentration of each of the predominant salt ions and summing to get the total. Effects of individual ions sometimes makes this measurement necessary. However, gravimetric determination in which a gravimetric or volumetric solution sample is evaporated and the mass of residuals determined is a much simpler procedure to measure the total salt concentration. The salinity is then given as parts per million (ppm = mg/kg) or milligrams per liter (mg/l).

An even easier and quicker method to get an estimation of the total salt concentration of a solution is to measure the electro-conductivity, EC, of the solution. The ions in a solution facilitate the flow of an electrical current through the solution. By inserting two electrodes in the solution and creating a voltage across the electrodes, the resistance to the flow of electricity (or inversely, the conductance) gives an indication on an equivalent basis, but because conductances are similar, and because most soils contain mixes of roughly the same ions, conductance can be used as a good estimate of total salt concentration. The dependability of EC salinity measurements is greatly increased by determining the relationship between EC and salt concentration for each mix of ions or group of measurements.

The standard electroconductivity cell is composed of two 1 cm^2 platinum plate spaces one centimeter apart. The resistance of the solution is determined by an alternating current wheatstone bridge. The resistance, R_s (ohms), is converted to conductance, EC, and usually expressed in millimhos or micromhos (μ mhos) per centimeter :

$$EC_T\ (millimhos) = \frac{1}{KR_s} \times 10^3 \tag{10.1a}$$

$$EC_T\ (micromhos) = \frac{1}{KR_S} \times 10^6 \tag{10.1b}$$

where K is the cell constant. Cell constants, which are often close to 1.0, should be printed on the cell, or can be determined by calibration (see Section 10.3).

Electroconductivity of solutions is strongly affected by the temperature, T (°C), of the solution. Consequently, the temperature must be determined and the measurement converted to EC at the standard temperature of 25°C. The effect of temperature on resistance is very nearly 2 percent per degree centigrade, and thus measurements can be converted to standard measurements by :

$$EC_{25^\circ} = \frac{EC_T}{1 + 0.02\,(T - 25^\circ)} \tag{10.2}$$

In this exercise, only total salt concentrations are measured. Although this is an important indicator, it does not measure the equally important although less common problems which arise from high concentrations of specific salt ions, such as sodium, magnesium and chloride. High proportions of exchangeable sodium ions, termed sodicity, cause soils to disperse and can result in very low infiltration rates. Alleviation of sodicity problems requires specific and often costly and time consuming reclamation and management practices. The problems with the second two ions are primarily physiological. If total salinity levels aren't sufficient to cause observed physiological problems, or specific ion problems are suspected, additional determinations of specific ion concentrations must be made. Methods are described in USDA Handbook #6 (USDA, 1954).

10.2 EQUIPMENT

1. Buechner or comparable funnel and filter paper
2. Vacuum flask with stopper, or vacuum extraction manifold and collection jars capable of safely containing 85 kPa (25 in. Hg) of vacuum
3. Crucible
4. Standard conductivity cell
5. Alternating current wheatstone bridge
6. Balance accurate to ± 1 mg
7. Crucible tongs
8. 600 ml beaker

Also : laboratory scoop, oven, vacuum system, dessicator, thermometer, #6 sieve

10.3 PROCEDURE

Note : Electroconductivity cells are normally provided with a cell constant, which is often 1.0. However, this constant should be periodically checked by dividing the measured EC_{25} (μmhos/cm) of a 0.1 N potassium chloride standard solution (0.7456 g of dry KCL per liter of distilled water) into 1411.8.

1. Prepare a saturated soil paste by adding distilled or deionized water to a sample of soil in the beaker (at least 300 g) while stirring with a laboratory scoop. The mixing process is easier if the sample is first passed through a #6 sieve. At saturation the soil paste glistens as it reflects light, flows slightly when the container is tipped, and slides freely and cleanly off the spatula for all but clayey soils.

2. Measure the saturation moisture content, W_S, of a subsample of the saturated paste by weighing the paste subsample wet and after drying in the oven for 24 hours.

3. Place a medium grade filter paper in the porcelain funnel and pour the saturated soil paste onto the paper. Make sure that the paper and bottom of the funnel are covered and that there are no vacuum leaks. Attach the porcelain funnel to the vacuum flask, as shown in Figure 10-1, or alternately to a vacuum extraction manifold (Fig. 10-2), and apply vacuum. Press and smooth the soil paste frequently to eliminate air pockets. Vacuum extraction should be terminated when air begins to pass through the filter and little water is being collected. If the extract is turbid, sediment has leaked through or around the filter and the sample must be refiltered.

4. Rinse the conductivity cell with the soil solution extract to be measured. The adequacy of rinsing is indicated by the absence of resistance changes with successive rinsings. If only a small amount of the soil solution extract is available, the cell may be rinsed with acetone and ventilated until it is dry.

5. Refill the cell with the soil solution extract and measure the resistance of the extract with the wheatstone bridge (Fig. 10.3), and the extract temperature, T, with a thermometer. Convert the R_S reading to a standard conductance reading using Eqs. 10-1 and 10-2.

6. Weigh a clean, dry crucible to the nearest 1 mg. Place the extract solution in the crucible and immediately weigh. Evaporate the extract in an oven. Remove the crucible and dry residuals from the oven, cool in a dessicator, and quickly weigh. Avoid touching a dry crucible with `moist' hands. Use tongs.

Figure 10.1 Extraction of a saturated extract with a Buechner funnel and vacuum flask.

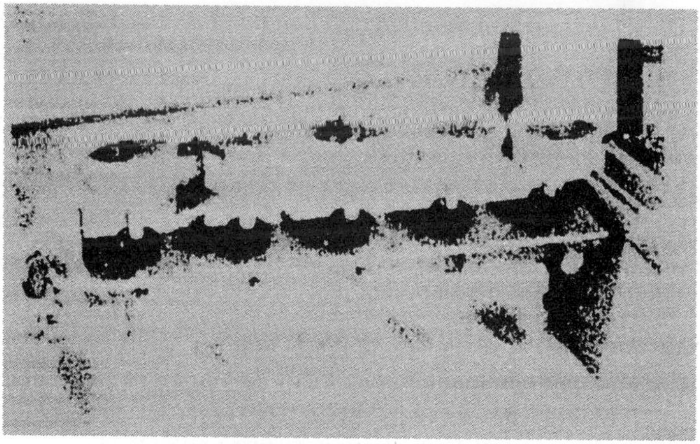

Figure 10.2 Extract collection with a vaccum manifold.

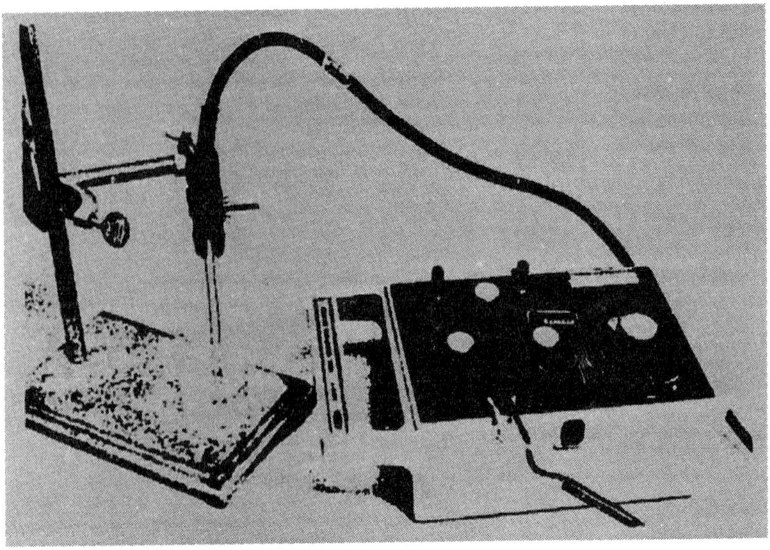

Figure 10.3 Measurement of extract resistance with a conductivity cell and Wheatstone bridge.

7. Calculate the salt concentration as the mass of the dry residuals divided by the solution mass and multiplied by 10^6.

8. Calculate the ratio of the extract salt concentration (ppm) to its electroconductivity (umhos/cm).

10.4 ANALYSIS AND DISCUSSION

The USDA Handbook #60, ``Diagnosis and Improvement of Saline and Alkali Soils" (USDA, 1954) gives an ``average" conversion factor of 0.64 to convert from EC_{25} (umhos/cm) to salt concentration (ppm). This number is often used, although the ratio can vary widely for different soils and waters. Would the use of the standard constant have caused significant error in the predicted salinity concentrations in the measured soils? Can a better constant be developed for the soils tested, or mushc each be evaluated independently?

Salt concentrations in the extracts depend upon the moisture content of the soil paste. Can variability in the measured concentrations (or EC) of the extracts from the same soil be related to variations in the estimated saturation moisture contents.

Handbook #60 lists the effect of soil salinity on crop growth as given in Table 10.1.

Table 10.1. Effect on crop growth on soil salinity as measured by the electroconductivity of a saturated soil extract (adapted from USDA, 1954).

SCALE OF CONDUCTIVITY, EC_{25} (umhos/cm)			
0	4,000	8,000	16,000
All crops thrive, no evidence of salt injury.	Sensitive crops do not thrive. Tolerant crops may do well.	Crop growth restricted yield usually poor.	Only a few species survive

According to Table 10-1, will any of the measured soils require special management to reduce the salt concentrations for optimum crop growth? What are the management procedures which could be followed ?

10.5 REFERENCES

Bureau of Reclamation (USBR) 1974. Earth Manual, Appendix E-8: Soluble Salts Determination of Soils. U.S. Department of Interior, USBR. Second Ed.

United States Department of Agriculture (USDA). 1954. Diagnosis and Improvement of Saline and Alkali Soils. L.A. Richards, ed. USDA Handbook No. 60.

Soil-Water Engineering Laboratory
Exercise 10
Soil Salinity Data Sheet

Name______________________________

Date______________________________

Sample Description 1.______________________________

2.______________________________

3.______________________________

Conductivity cell type and constant, K______________________________

Resistivity measurement device______________________________

	1	2	3
Electroconductivity:			
Extract resistance, R_s (ohms)			
Extract temperature, T (°C)			
Extract conductivity, EC_{25} (mmhos/cm)			
Gravimetric:			
Crucible mass, M_t (g)			
Extract and dish mass, M_{et} (g)			
Residue and dish mass, M_{rt} (g)			
Gravimetric salt concentration, C (ppm)[1]			
Ratio: C/EC_{25}			
Soil saturation percentage:			
Soil can mass, M_t (g)			
Saturated sample and can mass, M_{swt} (g)			
Oven dry sample and can mass, M_{st} (g)			
Saturation moisture content, Mass Basis, W_s[2]			

1 $C = \frac{M_{rt} - M_t}{M_{et} - M_t} \times 10^6$

2 $W_s = \frac{M_{swt} - M_{st}}{M_{st} - M_t} \times 100$

EXERCISE 11

WATER MEASUREMENT

11.1 INTRODUCTION

In order to efficiently manage irrigation water, the water must be measured. The measurement of open channel flows is usually of the flow rate or discharge, while volumetric measurements are often made of flows in pipelines. Flow rate measurements are required to match irrigation applications to soil intake rates and to design conveyance systems. Flow volumes, which can be determined by multiplying flow rates by time, are needed to determine total water applications and irrigation depths on a field.

Many type of flow measurement devices have been developed. In this exercise a few representative devices will be discussed and demonstrated. More thorough discussions of a wide variety of flow measurement structures are given in the Bureau of Reclamation's ``Wats" : Measurement Manual" (USBR, 1967); Section 15, Chapter 9 of the SCS National Engineering Handbook (SCS, 1962); and ``Discharge Measurement Structures", published by the International Institute of Land Reclamation and Improvement (ILRI, 1976). A more compete list of references is given at the end of this exercise.

Most flow measurement devices can be divided into two basic categories-those which measure the energy of a flow and those which measure the flow velocity. Examples of the first type include orifices, weirs, and flumes. Propeller meters, floats and the geometry of a jet of water freely flowing from vertical or horizontal pipes can be used to indicate flow velocities which can be related to discharge. Both types of measurements can be made in either closed conduits or open channels.

A review of basic fluid mechanics principles will aid in the understanding of flow measurement devices.

Fluid Dynamics of Flow Measurement

Flowing water has three types of energy :

1. Kinetic energy due to movement,
2. Potential energy due to pressure, and
3. Potential energy due to elevation.

These energy forms, when added together, give the total energy, or when expressed per unit force, the head, H_t (m), of the flowing water.

$$H_t = \frac{V^2}{2g} + \frac{p}{\rho g} + H, \tag{11.1}$$

where :

V = flow velocity (m/sec),

ρ = fluid density (kg/m^3),

g = acceleration of gravity (9.80 m/sec^2),

p = pressure (Pa), and

H = elevation above a datum (m).

The total energy or head of a flow between two points along a flow path is decreased by friction and form losses, H_L. The energy balance between the two points can thus be expressed by Bernoulli's Equation :

$$\frac{V_1^2}{2g} + \frac{p_1}{\rho g} + H_1 = \frac{V_2^2}{2g} + \frac{p_2}{\rho g} + H_2 + H_L. \tag{11.2}$$

Figure 11.1 is a definition sketch for the energy of water flowing through an orifice. The sum of the elevation head H, and pressure head, $\frac{p}{\rho g}$, is often termed the hydraulic head, h. Snce the pressure on all sides of the jet of water below the oriice is atmospheric (zero), if we move the datum to the centerline of the orifice, the downstream hydraulic head will be zero, and Bernoullis' Equation simplifies to :

$$\frac{V_1^2}{2g} + h_1 = \frac{V_2^2}{2g} + H_L. \tag{113}$$

Solving for V_2 :

$$V_2 = \left[2g \left(\frac{V_1^2}{2g} + h_1 - H_L \right) \right]^{1/2} \tag{11.4}$$

In most cases, the upstream velocity head, $\frac{V_1^2}{2g}$, and the head loss of the contracting flow, H_L, will be small relative to the hydraulic head, h_1, so Eq. 11.4 can be simplified to :

$$V_2 = \sqrt{2gh_1} \,. \tag{11.5}$$

The velocity can be converted to a flow rate, Q (m^3/sec), using the continuity equation, $Q_1 = Q_2$, and the relationship Q = VA, giving :

$$Q = A\sqrt{2gh_1} \,. \tag{11.6a}$$

where A is the cross-sectional area of the orifice.

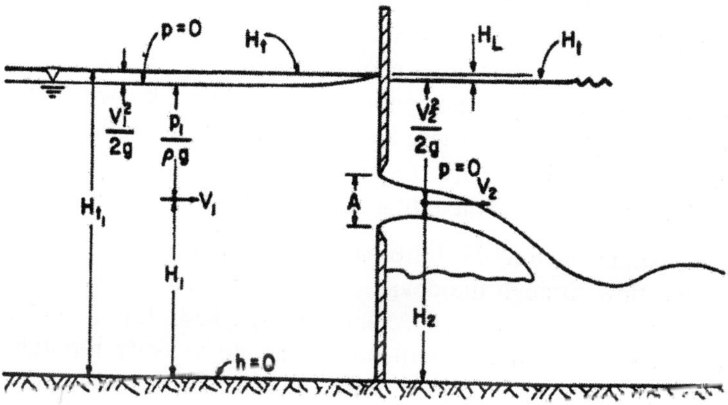

Figure 11.1. Flow through an orifice.

Because the converging streamlines will cause the jet to contract smaller than the orifice size, A, there will be some head loss through the constriction, and the velocity distribution effects have been ignored, the actual discharge will be less than this predicted ideal discharge. A discharge coefficient, C_d, is used to correct for these factors.

$$Q = C_d \, A \, \sqrt{2gh_1} \tag{11.6b}$$

By similar analyses, expressions for the flow rate over weirs or through flumes as function of hydraulic head can be derived. In all cases, a discharge coefficient must be used to correct for deviations from asumptions and flow contraction. Because the flow area, A, through open channel devices will also be a function of the upstream water depth or head, the discharge equation exponent for open channel devices will be greater than the 0.5 for orifices. Rectangular weirs and flumes have a theoretical exponent of 1.5, while V notch weirs have a thoretical exponent of 2.5. The exponents for some flow measurement flumes, because of their specific upstream geometrics, will vary from these values which are based only on the cross-sectional shape at the contraction. However, most devices or structures through which the flow is contracted and accelerated to at least critical flow velocity (called ``free flow") will follow the basic equation :

$$Q = Ch_1^{\mu} \tag{11.7}$$

The coefficient and exponent for several standard flow measurement devices have been determined in laboratory studies and are available from many sources, such as those in the listed references. Other structures can be calibrated in the field or laboratory by measuring the upstream head while a known flow rate is passing through them. These head discharge data can then be fit to Equation 11.7 graphically on log-log paper or by regression techniques to determine the rating curve for the device.

Equations such as 11.7 are based on two important conditions. First, the flow through the device contraction must reach or exceed critical velocity. If critical velocity is reached, flow conditions downstream of the device will not affect the velocity through the contraction. Devices with constant flow cross-sections such as orifices, as well as other devices if specially calibrated, can also be used under submerged flow conditions (when critical velocity is not reached) if the dowstream head, h_2 is also measured. The existence of a hydraulic jump downstream of the constriction always indicates free (unsubmerged) flow.

Equation 11.7 is also based on the approximation that the velocity upstream of a device is zero. This is obviously never true, but since velocity head (see Eq. 11.4) is a function of velocity squared, the approximation is often good. If the velocity head upstream of the device is greater than 1 or 2 percent of the hydraulic head, h_1, then this head must be added to h_1 to obtain the total head. This is equivalent to V_1 being larger than about $V_2/8$ or $A_1>8A_2$. Corrections for approach velocities are alternatively sometimes made with a velocity coefficient, C_v. Flumes, in which the head measurement is made in a fixed contracting section, have been calibrated under these specific approach conditions and the discharge equations are not sensitive to approach velocity velocity with given limits.

Velocity Measurement Devices

Several flow measurement devices have been developed which measure the velocity of flow. If the average flow velocity, $\overline{V}$, can be determined at a point of known cross-sectional area, A, the flow rate can be determined by :

$$Q = \overline{V} A \tag{11.8}$$

Propellor and vane meters utilize the momentum of flowing water to deflect vanes. The vane deflection or propellor rotation is then calibrated in terms of water velocity. Because the flow velocity varies across a flow cross section, the device must (a) be calibrated for a given position in a given cross section, (b) be installed such that it intercepts streamlines flowing at the average velocity, or (c) be moved aroung to determine average velocity.

Floats such as wooden objects can be placed in steady, uniformly flowing reaches of channels to indicate the flow velocity. The float must be calibrated in a representative channel, or designed to be carried along at the average velocity. A slug of dye can also be injected into a stream to determine the flow velocity.

When water exits from a pipe, its momentum will carry it a certain distance upward or horizontally before it falls to the ground. This distance can be related, using momentum principles, to exit velocity. However, because of the nature of water jets, empirically derived rating curves will given more accurate estimates of discharge than calculations based on theoretical principles.

Selected Flow Measurement Structures

Orifices

Figures 11.2 shows an orifice plate installed in an open channel. If the orifice jet is submerged as shown, the total head loss or drop in the water surface is less than that for a free flowing jet. The illustrated orifice is fully contracted, i.e., the perimeter of the opening is far enough from the walls of the approach channel and water surface that the jet fully contracts when passing through the opening. Orifices can be of various shapes with circular and rectangular being the most common.

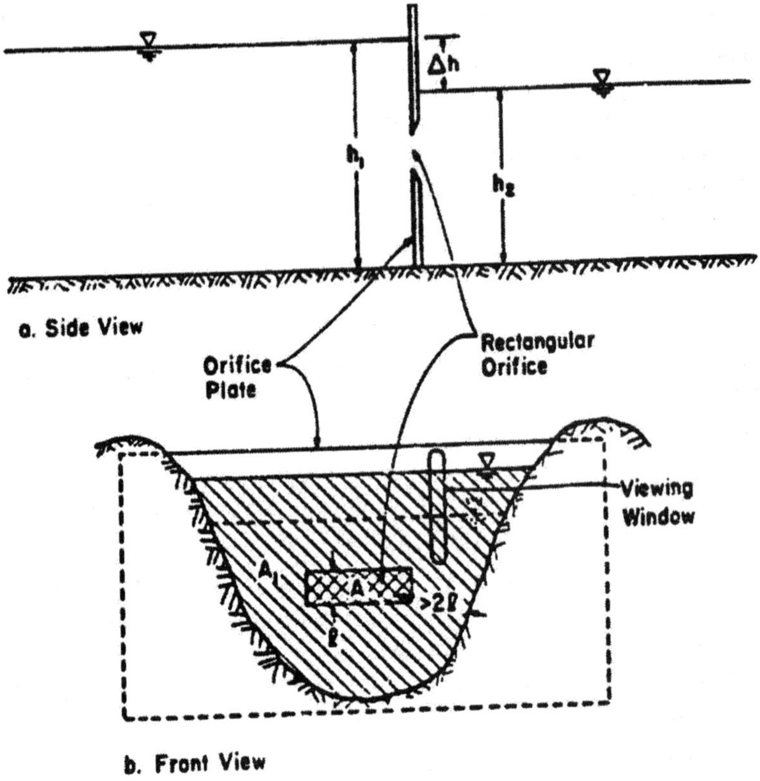

Figure 11.2. Rectangular submerged orifice.

A standard orifice must :

1. be placed perpendicular to the flow,
2. have sharp edges, (no more than 1-2 mm thick),
3. be contracted from the channel boundary and water surface at least twice the smallest opening dimension (l in Fig. 11.2), and
4. have a cross-sectional flow area upstream of the orifice, A_1, at least 8 times the orifice cross-sectional area, A.

The discharge equation for a standard submerged orifice is Eq. 11.6b with an orifice coefficient of 0.61. The head is the difference between the upstream and downstream water levels, (h_1–h_2).

$$Q = 0.61\,A\sqrt{2g\,(h_1 - h_2)} \tag{11.6c}$$

If the upstream channel cross-sectional area, A_1, is not 8 times larger than the orifice opening area, A, then the approach velocity must be taken into consideration. the upstream velocity head, h_v, given by :

$$h_v = \frac{V_1^2}{2g} = \frac{Q^2}{2gA_1^2} \tag{11.9}$$

must be added to the upstream hydraulic head, so that :

$$Q = 0.61\,A\sqrt{2g\,(h_1 + h_v - h_2)}. \tag{11.10}$$

If a portion of the orifice is suppressed, (i.e., an edge of the orifice does not protrude from, but is even with, the upstream channel floor or walls) the discharge coefficient, C_d, must be increased by 1.5 percent for every 10 percent of the orifice perimeter which is suppressed.

$$C_d' = 0.61\,(1 + 0.15\,r), \tag{11.11}$$

where :

C_d' = the coefficient for a partially suppressed orifice, and

r = the ratio of the suppressed to the contracted portion of the orifice perimeter.

A common example of a suppressed orifice is a sluice gate in which the bottom and sides are suppressed and the top is contracted. An orifice which is neither fully contracted nor suppressed along any part of its perimeter must be caslibrated to determined the discharge coefficient.

If an orifice's edges are not sufficiently sharp to create full contraction of the jet, the coefficient will be higher than 0.61 and may be as high as 0.80. Water passing through such orifices will `cling' to the edges of the opening rather than springing free. Such orifices must be calibrated.

The head loss across a submerged orifice, (h_1 – h_2) or Δh, can be easily determined by constructing a window in the plate through which the upstream level can be viewed from the downstream side and the difference measured with a rule or staff gage.

Orifices are also used in pipe conveyance systems as shown in Figures 11.3 and 11.4. In such applications, the jet is generally not fully contracted and the velocity of approach must be considered. Figure 11.5 shows the variation in the orifice discharge coefficient, C_d, with the degree of contraction represented as the ratio of the orifice diameter, d, to pipe diameter, D. This coefficient should then be used in place of 0.61 in Equation 11.10. Also, with circular pipes and orifices, h_v can be derived in terms of diameters so that :

$$Q = C_d\, A\sqrt{2g\,(\Delta h + h_v)} = \frac{C_d A}{\sqrt{1 - C_d^2\left(\frac{d}{D}\right)^4}}\sqrt{2g\,\Delta h}\ . \qquad (11.10a)$$

Such orifices can be placed in pipelines or on the outflow ends of pipes. Heads can be measured with piezometer taps a distance D upstream from the orifice plate and about 0.5D downstream from the plate. The h_2 value for pipe end orifices is atmospheric, or zero.

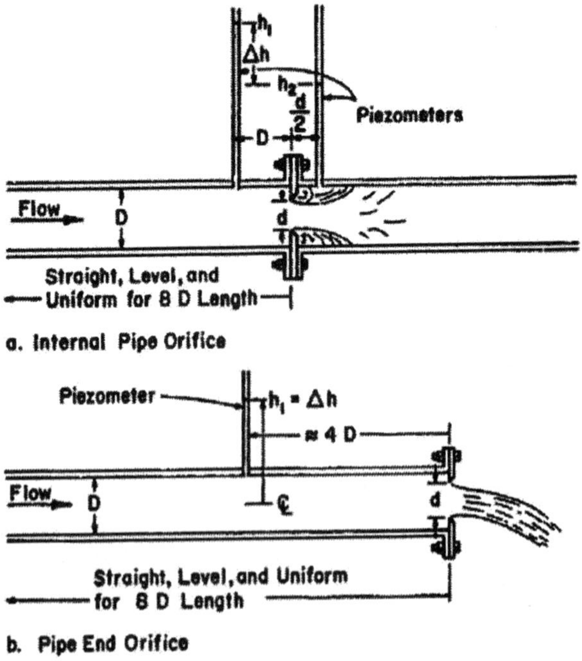

Figure 11.3. Closed conduit concentric orifices.

Figure 11.4. Measurements being taken from manometers attached to an internal pipe orifice.

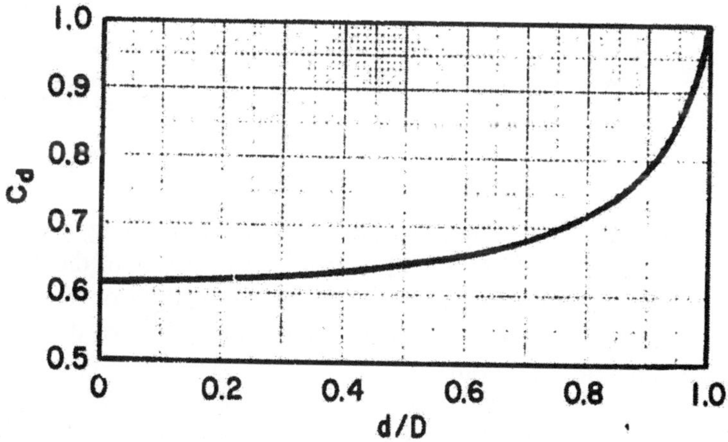

Figure 11.5. Orifice discharge coefficient, C_d, for a concentric circular orifice of diameter, d, in a pipe of diameter, D (from Albertson *et al.*, 1960).

A primary disadvantage of orifices in pipelines is the head loss and the resultant higher energy costs to maintain a required pressure in the pipeline. Consequently, orifice-type devices in which most of the velocity head is recoverted to pressure head, such as venturis, are sometimes used.

Spiles and siphon tubes also act like orifices. In short lengths of pipe, the small additional head looses due to pipe friction (H_L in Eq. 10.2) are also proportional to the velocity head. Therefore, their effect is to reduce the discharge coefficient slightly. Spiles and siphon tubes can therefore be used as flow measurement devices if they are calibrated. Their discharge will also follow the form of Eq. 11.6c, but the discharge coefficient will vary depending upon the shape and length of the tube and the entrance and exit conditions. Coefficients close to 0.60 are common.

Weirs

Weirs are like free flowing (unsubmerged) orifices which are not full but have an open water surface. Therefore, as the head increases, the flow depth over the weir and thus the area, A, increases. Utilizing Bernoulli's equation (Eq. 11.2) between the upstream channel and the weir (see Fig. 11.6), and making the same assumptions as were made with the orifice (H_L and $\frac{V_1^2}{2g}$ are zero) results in :

$$h_1 = \frac{V_2^2}{2g} + h_2 \tag{11.2a}$$

or :

$$V_2 = \sqrt{2g\left(h_1 - h_2\right)} \tag{11.12}$$

By continuity, $Q_1 = V_2A$; and for a rectangular opening, such as that shown in Figure 11.6, $A = Lh_2$ where L is the width of the weir. Also, the flow depth will pass through critical depth at or near the weir, and critical depth in a rectangular channel is about two-thirds of the upstream head, or $(2/3)h_1$, so :

$$Q = Lh_2\sqrt{2g\left(h_1 - h_2\right)} \tag{11.13}$$

$$= L\left(2/3\right)h_1\sqrt{2g\left(1/3\right)h_1},$$

or, $$Q = C_d \left(2/3\right) L \sqrt{\left(2/3\right) g}\ h_1^{3/2} \tag{11.14}$$

Thus, the exponent, μ (Eq. 11.7) for rectangular weirs will be 3/2, as compared to 1/2 for orifices. The exponent value will vary with the cross-sectional shape of the weir (i.e., rectangular, V notch, or trapezoidal). The discharge coefficient which, as with orifices, is required to correct for the fact that water is not an ``ideal" fluid, will vary with both cross-sectional shape and type (i.e., suppressed or contracted, sharp, short, or broad crested). Table 11.1 shows a rectangular, contracted sharp-crested weir in use.

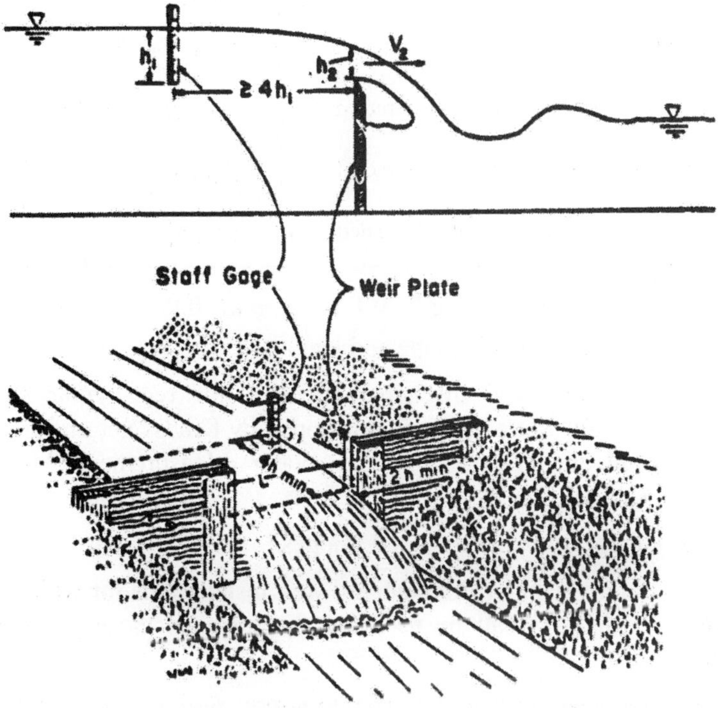

Figure 11.6. Sharp-crested contracted rectangular weir (from SCS, 1962).

The edges of a contracted weir, like those of a contracted orifice, must be at least twice the head on the weir from the sides or bottom of the channel. A suppressed weir has no side contractions-only a bottom sill. Sharp crested weirs, like orifices, are thin enough at the crest that the water ``springs free" of the weir. A weir plate tapered to 1 to 2 mm at the crest is sharp crested. Broad-crested weirs have horizontal and prismatic crests long enough that the flow streamlines become parallel while still on the crest. Crest lengths should be between twice and twelve times the head on the weir, h. The crest lengths of short-crested weirs are between these two categories.

Table 11.1. Discharge Equations for Common Weirs.

Shape	Type	Discharge Equation Q in m^3/sec, L and h_1 in m
Rectangular	Sharp crested, suppressed	$Q = CLh_1^{3/2}$ (1)
Rectangular	Sharp crested, suppressed	$Q = 1.84\ Lh_1^{3/2}$ (2)
Rectangular	Sharp crested, contracted	$Q = 1.84\ (L - 0.2\ h)h_1^{3/2}$ (2)
Cipolletti(3)	Sharp crested, contracted	$Q = 1.86\ Lh_1^{3/2}$
90° V-Notch	Sharp crested, contracted	$Q = 1.34\, h_1^{2.48}$

(1) $C = C_d\ (2/3)\sqrt{2g/3} = 1.70C_d \quad with\ 0.9 < C_d < 1.05$ depending on still shape and lenght.

(2) An improved discharge equation for rectangular sharp crested weirs has been developed by Kindsvater and Carter and is presented in the Water Measurement Manual (USBR, 1967).

(3) Trapezoidal with 1:4 side slopes.

Conditions required for standard weir installations for which the discharge equations can be used are the same as those given for orifices, plus :

1. the head on a weir, h, should not be more than one-third of the width, L;
2. head measurements should be made at least $4h_1$ upstream of the weir with the weir crest as the datum;
3. an air space should exist under and around the jet to ensure free flow; and
4. suppressed weirs must must be aerated below the jet to prevent creation of a vacuum.

Figure 11.7. A portable rectangular, contracted, shapr-crested weir used in an earthen channel.

Weirs are not recommended under submerged flow conditions.

As with orifices, if the velocity of approach is greater than one-eighth the flow velocity over the weir, a correction should be made in the upstream head value, h_1 :

$$h_1' = \left[(h_1 + h_\upsilon)^\mu \quad h_1^\mu\right]^{1/\mu}, \tag{11.15}$$

where h_1' is the head corrected for velocity of approach.

Flumes

Flow measurement flumes are special types of broad - or short-crested weirs which include an entrance and exit section. Their discharge equations are therefore similar to weir equations. Flumes have been developed primarily to reduce device head losses, enhance portability, and improve passage of sediment and trash that is common in open channel conveyance systems.

Because the upstream head, h_a, is measured in the converging section of most flumes and because of friction losses along the flume walls and floor, they must be calibrated in the laboratroy. Any flume which uses standard calibrations must then be a dimensional duplicate of the original. Because flumes include an entrance and exit section and head measuring gages, the only requirements for proper installation is to ensure they are level, the inflow to the flume is below a specified velocity and relatively uniformly distributed across the width, and the flow depth and degree of submergence are within rated ranges.

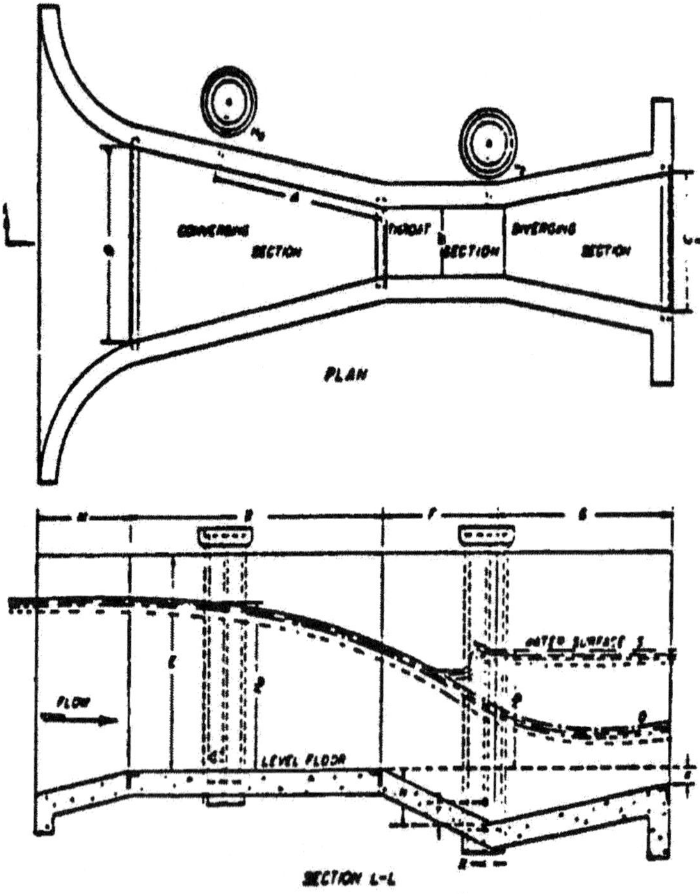

Figure 11.8. Parshall flume shape and dimensions showing stilling wells for h_a and h_b measurments (from SCS, 1962).

The most commonly used flume in the world today is the Parshall flume, developed at Colorado State University by Ralph Parshall. Figure 11.8 shows the important dimensions of small Parshall flumes. Parshall flume sizes are based on the throat width, W. The throat section has parallel vertical walls with a sloping floor. Free flow conditions occur in small Parshall flumes at submergence ratio, h_b/h_a, below 0.5 to 0.6. The flumes have been calibrated under submerged flow conditions with the downstream head measureent, h_b, made in the throat or tail section. Table 11.2 shows the dimensions of small Parshall flumes and the free flow discharge equations.

Table 11.2. Dimensions for small and intermediate size Parshall flumes (see Fig. 11.8).

Flume Size	W (m)	A (m)	B (m)	C (m)	D (m)	E (m)	F (m)	G (m)	[illegible] (m)	[illegible] (m)	X (m)	Y (m)	Free Flow Range (m^3/sec)	Discharge Equation (m^3/sec) - (m)	Transition Submergence h_b/h_a
1 in	.025	.242	.356	.093	.167	.229	.076	.203	.019	.029	.008	.013	.0003 - .006	$Q = .0604 h_a^{1.55}$	[illegible]
2 in.	.051	.276	.406	.135	.214	.254	.114	.254	.022	.043	.016	.025	.0006 - .014	$Q = .1207 h_a^{1.55}$	[illegible]
3 in	.076	.311	.457	.178	.259	.457	.152	.305	.025	.057	.025	.038	.0008 - .028	$Q = .1771 h_a^{1.55}$	[illegible]
6 in	.152	.414	.610	.394	.397	.610	.305	.610	.076	.114	.051	.076	.0015 - .110	$Q = .3812 h_a^{1.58}$	[illegible]
9 in	.229	.587	.864	.381	.575	.762	.610	.914	.076	.114	.051	.076	.0025 - .250	$Q = .5354 h_a^{1.53}$	[illegible]
1 ft.	.305	.914	1.343	.610	.845	.914	.610	.914	.076	.229	.051	.076	.0031 - .450	$Q = .6909 h_a^{1.52}$	[illegible]
1 1/2 ft	.457	.965	1.419	.762	1.026	.914	.610	.914	.076	.229	.051	.076	.0042 - .680	$Q = 1.056 h_a^{1.54}$	[illegible]
2 ft	.610	1.016	1.495	.914	1.207	.914	.610	.914	.076	.229	.051	.076	.0119 - .937	$Q = 1.428 h_a^{1.55}$	[illegible]
3 ft	.914	1.118	1.645	1.219	1.572	.914	.610	.914	.076	.229	.051	.076	.0173 - 1.430	$Q = 2.184 h_a^{1.57}$	[illegible]

In order to make installation in sloping-sided flat bottomed earthen or lined channels easier, trapezoidal long-throated flumes have been developed and calibrated. These trapezoidal flumes are also referred to as WSC flumes due to early work done at Washington State College (now University). As shown in Figures 11.9, trapezoidal flumes have flat bottoms and contracted, sloping side walls. Dimensions of commonly used small trapezoidal flumes are given in Table 11.3.

Because of their complex shape, the discharge equations for trapezoidal flumes are not as simple as with rectangular flumes. Discharge equations for sizes commonly used in irrigation channels are given in Table 11.3. Two advantages of trapezoidal flumes, in addition to their being easy to install, is that the free flow discharge equations are valid for submergence levels up to 80 percent, and, due to the sloping walls, the range of flows which can be measured in a given flume is large.

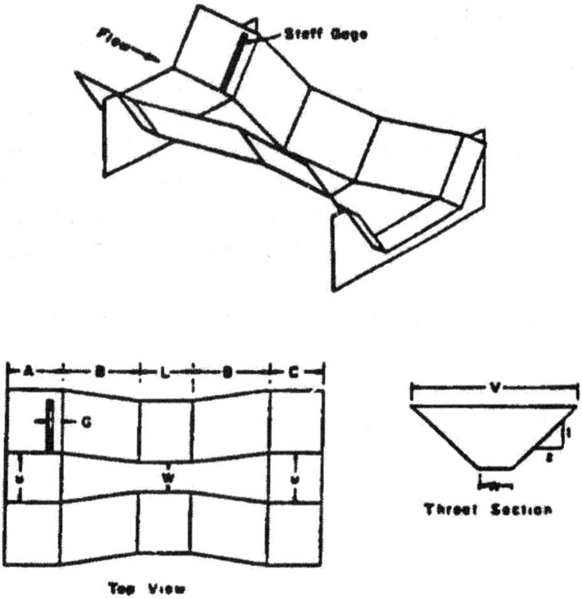

Figure 11.9. Trapezoidal flume shape and dimensions.

In order to simplify flume construction, flumes with no throat, termed Cutthroat flumes, were developed by G.V. Skogerboe and associates at Utah State University. The flume, shown in Figures 11.10 and 11.11, has a flat bottom and vertical sided converging and diverging sections. The flume dimensions have been proportioned such that any length or width of flume can be used.

Free flow discharge equations for Cutthroat flumes are also of the form of Equation 11.7. The exponent, μ, is a function of the flume length while the coefficient, K, is a function of both length and width.

$$Q = KW^{1.025}\ h_a^{n_1}, \tag{11.16}$$

where :

Q = discharge (m^3/sec),

W = flume width at the throat (m),

h_a = the upstream head measurement (m), and

n_1 and K = coefficient dependent on flume length.

Values for K and n_1 can be taken from Figure 11-12.

Table 11.3. Dimensions, flow ranges and discharge equations for six trapezoidal flumes.

Flume No.	Dimensions (m) (see Fig. 11-9) z	W	U	V	L	A	B	C	G	Flow Range (m^3/sec)
1[(1)]	1.0	.122	.305	.935	.305	.381	.433	.152	.046	.001 - .170
2[(1)]	1.25	.305	.610	2.590	.760	.910	.910	.460	.150	.015 - 1.60
3	0.58	.203	.406	1.017	.508	.457	.498	.152	.076	.007 - .50
4	1.0	.051	.124	.589	.216	.203	.213	.076	.038	.001 - .04
5	0.58	.000	.051	.197	.178	.178	.176	.076	.038	.0003 - .007

DISCHARGE EQUATIONS[(2)]
(Free Flow)

Flume No.	Discharge Equations (Q in m^3/sec and h in vertical meters)
1	$Q = 1.783h_a^{2.5} + 0.106h_a^{1.5} + 0.002$
2	$Q = 1.810h_a^{2.5} - 1.359h_a^{2} + 1.325h_a^{1.5} - 0.244h_a + 0.036h_a^{1/2} - 0.002$
3	$Q = 0.806h_a^{2.5} + 0.374h_a^{1.5}$
4	$Q = 1.795h_a^{2.5} + 0.056h_a^{1.5}$
5	$Q = 0.655h_a^{2.46}$

(1) Sized to fit into standard slip-form concrete lined channels.

(2) Equations for flumes 1 through 5 are empirically derived from calibration data.

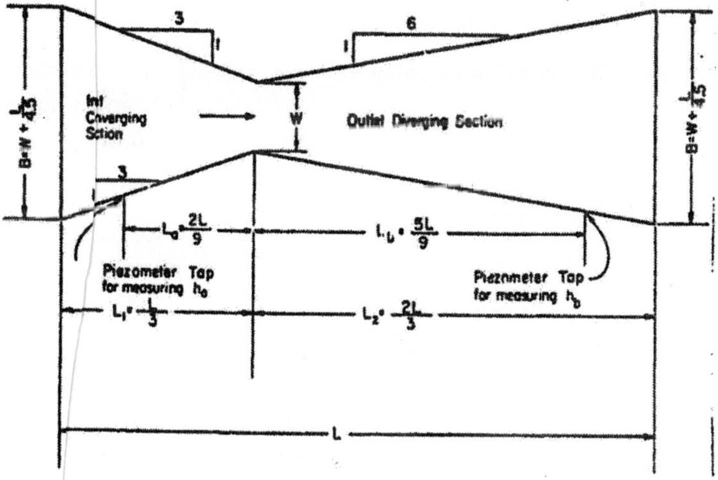

Figure 11.10. Cutthroat flume dimensions (from Skogerboe *et al.*, 1974).

Figure 11.11. A cutthroat flume being installed and leveled.

Cutthroat flumes have also been calibrated under submerged flow conditions. Submergence of shorter flumes occur in the range of 60 to 70 percent. The submerged flow discharge equation is :

$$Q = \frac{K_1\ W^{1.025}\ (h_a - h_b)^{n_1}}{(-\log S)^{n_2}}, \tag{11.17}$$

where

h_b = the downstream depth (m),

S = the submergence (h_b/h_a), and

K_1 and n_2 = coefficients dependent on flume length.

Values for K_1 and n_2 can also be determined from Figure 11-12, as can the transition submergence, S_t, at which flow conditions change from free to submerged.

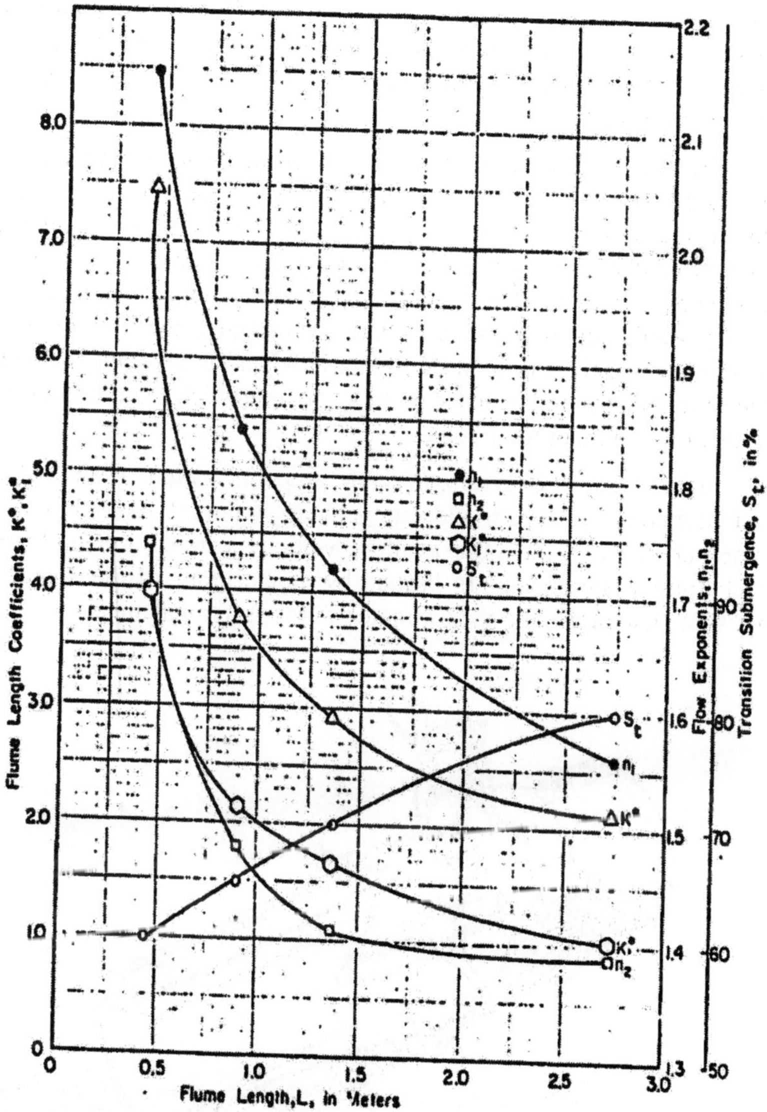

Figure 11.12. Generalized free flow and submerged flow coefficients and exponents and S_t for cutthroat flumes (from Skogerboe *et al.,* 1973).

Propeller Meters

Propeller or current meters are propellers on shafts which, when inserted into flowing water, will turn. The rotational velocity of the propeller is a function of the flow velocity. Thus, the frequency of rotation or accumulated number of revolutions can be related to the flow velocity or volume of water passing the propeller in a period of time, respectively.

Propeller meters used in small conveyance systems are usually calibrated for a specific cross-sectional area so that the intercepted flow velocity can be converted directly to a discharge rate or volume. Such propeller meters are either inserted into pipelines of specificed diameters as shown in Figure 11-13, or are permanently mounted in a short portable pipe section which is then attached to the end of a pipe or, less commonly, installed through a bulkhead in an open channel. Such propeller meters usually include gauges which read directly in terms of flow rate or total volume.

In large canals or rivers, portable propeller meters, often called current meters, are used to indicate velocity at a point in the flow. By sampling many points, the average velocity can be found. Such discharge measurements are limited by the accuracy of the velocity averaging technique and the estimate of cross-sectional area.

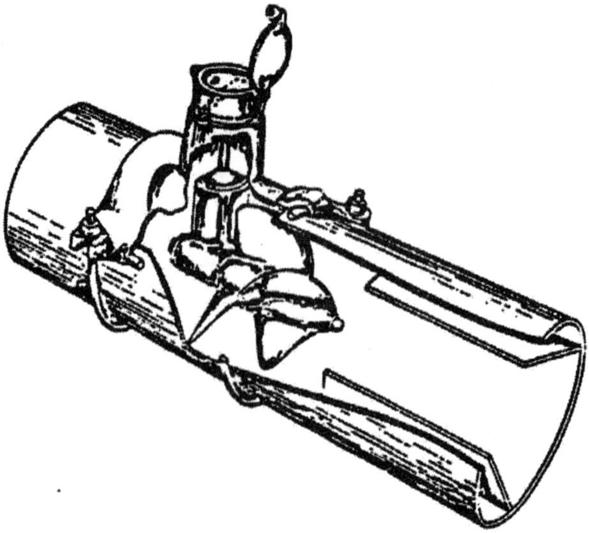

Figure 11-13. Propeller meter installed in a pipeline (from SCS, 1962).

Propeller meters cause very little head loss are thus energy efficient in pipeplines. However, they will catch trash in the water and must be checked often, or used only in clean water. Also, as with any mechanical device, they are subject to weat and must be maintained.

11.2 EQUIPMENT

1. Orifice plate installed in a bulkhead for open channels
2. Orifice plate for the delivery pipeline including piezometers
3. Sharp-crested weir (rectangular, Cipoletti, or V notch)
4. Parshall, trapezoidal, or Cutthroat flume
5. Propeller meter
6. Several siphon tubes of the same shape and diameter
7. Carpenter's level
8. Staff gauge
9. Small diameter clear flexible tubing (about 3 m)
10. Plastic sheeting for erosion control
11. Open channel supplied with 0.02 to 0.06 m^3/sec of water from a pipe.

Also : shovels, watch, and meter stick

11.3 PROCEDURE

1. Install the orifice plate in or on the end of the delivery pipeline. Use the orifice as the standard against which other devices are compared. If delivery is from an open source, use in orifice installed in the open channel as the standard.
2. Install the remaining devices in the channel ensuring that the weir and orifice plates are vertical, perpendicular to the flow, and fully contracted; and the flumes are level in both directions. The standard entrance and exit conditions must also be met.
3. Measure flows through each device at three flow rates.

4. Calculate the fow rates using the equations given in the Introduction section and compare with the discharge measured through the orifice. Calculate the percentage error in the measurement by the formula :

$$\left[\frac{Q_{device} - Q_{s\tan dard}}{Q_{s\tan dard}}\right] \times 100 \qquad (11.18)$$

5. Set sufficient siphon tubes over a dam or bulkhead installed across the channel to carry the maximum discharge with about 30 cm of head loss (Δh). The downstream end of the tubes (with the clear tubing used as a piezometer) at several discharge rates (measured by the orifice). Determine the discharge equation for the siphons by plotting Δh vs. Q/n, where n = the number of tubes, on log-log paper. A linear plot would indicate the siphons follow Equation 11.7, the slope would be the exponent, μ, and the intercept the coefficient, $C = C_d A \sqrt{2g}$, where A is the cross-sectional area of each tube. Calculate the discharge coefficient for the siphon tubes.

11.4 ANALYSIS AND DISCUSSION

Each flow measurement device has certain characteristics which influence its use and desirability under different conditions. These characteristics include :

1. Cost
2. Accuracy
3. Ease of use
4. Ease of installation
5. Required maintenance
6. Required maintenance
7. Portability
8. Potential sediment or trash interference
9. Adaptability to automated switches or recording devices

Rate each of the devices for each of the these characteristics. Indicate which device or devices you would install on your farm (real or hypothetical) and why.

11.5 REFERENCES

I. Orifices

1. Albertson, M.L., J.R. Barton, and D.B. Simons, 1960. Fluid Mechanics for Engineers. Prentice-Hall. Englewood Cliffs, N.J.
2. American Society of Mechanical Engineers (ASME). 1959. Fluid meters, their theory and application (5th Edition). ASME, 29 W 33rd Street, New York, New York.
3. International Institute for Land Reclamation and Improvement (ILRI). 1976. Discharge measurement structures. ILRI Publication No. 20, Wageningen, The Netherlands.
4. United States Bureau of Reclamation (USBR). 1967. Water measurement manual (2nd Edition), Denver, Colorado. (metric supplement was printed in 1971).
5. U.S. Soil Conservation Service (SCS). 1962. Measurement of irrigation water. SCS National Engineering Handbook, Section 15, Chapter 9, USDA.

II. Weirs

A. Sharp Crested

1. USBR, 1967
2. SCS, 1962
3. ILRI, 1976

B. Broad Crested

1. ILRI, 1976
2. Replogle, John A. and Albert J. Clemmens. 1978. Measuring flumes of simplified construction. Amer. Soc. of Agri. Engr. Paper No. 78-2506.

III. Parshall Flumes

1. USBR, 1967

2. SCS, 1962

3. ILRI, 1967

IV. Trapezoidal Flumes

1. SCS, 1961.

2. American Society of Agricultural Engineers (ASAE). 1980. Trapezoidal flumes for irrigation flow measurement, ASAE Standard 5359.1. In Agricultural Engineer's Handbook, 1980-81, St. Joseph, Michigan.

3. A.R. Robinson and A.R. Chamberlain. 1960. Trapezoidal flumes for open-channel flow measurement. Trans ASAE, Vol. 3, No. 2.

V. Cutthroat Flumes

1. Skogerboe, G.V., R.S. Bennett and W.R. Walker, 1973. Selection and installation of cutthroat flumes for measuring irrigation and drainage water. CSU Experiment Station Tech. Bulletin No. 120, Fort Collins, Colorado; or, by the same authors : Installation and field use of cutthroat flumes for water management. Water Management Tech. Report No. 19, CSU (1972).

Soil-Water Engineering Laboratory
Exercise 11
Water Measurement Data Sheet

Name____________________

Date____________________

Location____________________

Water Source____________________

I. **Standard Device**

Type of device____________________

Dimensions (m)____________________

Pipe or Channel Dimensions (m)____________________

Discharge Equation____________________

II. **Submerged Orifice**

Orifice Dimensions (m)____________________

Cross-Sectional (C.S.) Area, A (m^2)____________________

Contraction____________________

Upstream C.S. Area, A_1 (m^2)____________________

A_1/A____________________

Discharge Equation____________________

Discharge Measurements

Standard Device				Submerged Orifice			
h_1 (m)	h_2 (m)	Δh (m)	Q (m^3/sec)	h_1 (m)	h_2 (m)	Δh (m)	Q (m^3/sec)

Comments____________________

Soil-Water Engineering Laboratory
Exercise 11
Water Measurement Data Sheet (cont'd)

III. Sharp-Crested Weir

Weir type ____________________

Weir dimensions: Width, L (m)__________; Side Slope, z__________

Maximum allowable head, h_{max}, (m)____________________

Contraction____________________

A_1/A at h_{max}____________________

Discharge Equation____________________

Discharge Measurements

Standard Device				Weir	
h_1 (m)	h_2 (m)	Δh (m)	Q (m^3/sec)	h_1 (m)	Q (m^3/sec)

Comments____________________

____

IV. Flume

Flume type____________________

Flume size: Width (m)______; Length (m)______; Side Slope ______

Maximum allowed submergence for free flow, S_t____________________

Discharge Equation____________________

Discharge Measurements

Standard Device				Flume			
h_1 (m)	h_2 (m)	Δh (m)	Q (m^3/sec)	h_a (m)	h_b (m)	S, (h_b/h_a)	Q (m^3/sec)

Comments____________________

____

Soil-Water Engineering Laboratory
Exercise 11
Water Measurement Data Sheet (cont'd)

V. Propeller meter

Meter size (diameter)____________________________

Discharge Measurements

Standard Device				Propeller meter		
h_1 (m)	h_2 (m)	Δh (m)	Q (m^3/sec)	Volume (units)	Time (min)	Q (m^3/sec)

Comments____________________________________

____

VI. Siphon tube calibration

Tube diameter (m)____________________________

Tube length (m)____________________________

Tube C.S. area (m^2)____________________________

Number of tubes used, n____________________________

Siphon Tube Calibration

Standard Device				Siphon Tube			
h_1 (m)	h_2 (m)	Δh (m)	Q (m^3/sec)	h_1 (m)	h_2 (m)	Δh (m)	Q/n (m^3/sec)

Discharge Coefficient, C_d ____________________________

Exponent, μ ____________________________

Comments____________________________________

____

EXERCISE 12

MEASURING WEATHER PARAMETERS

12.1 INTRODUCTION

Previous exercises have discussed methods to measure soil and water factors which affect the water supply to plant roots. However, it is the atmosphere in which a plant grows which determines how much water the plant needs to grow and produce at its potential. Although a small percentage of the water a plant takes up is used for cellular growth, most is transpired into the atmosphere. The rate of plant transpiration and evaporation from the soil and plant surfaces is affected by the energy status of its surrounding atmosphere, as indicated by the weather parameters of solar radiation, temperature, humidity, and wind run. In addition to these weather parameters which influence crop water use, precipitation which enters the soil serves as a source of water.

By estimating from weather measurements how much water a crop has evapotranspired, and how much precipitation has replenished the soil water reservoir, the changes in the soil water storage can be calculated and the need for supplemental water determined. This process is called irrigation scheduling. Irrigation scheduling methods are described in several of the listed references. Evapotranspiration and precipitation data are also required to estimate peak and seasonal crop water requirements for the planning of irrigation projects.

Crop water use is determined by the potential of the atmopshere to vaporize liquid water and by the availability of water at or near the plant and soil surface. The atmospheric potential to vaporize water from plants and moist soils is called potential evapotranspiration, E_{tp}. Crop water use or consumptive use, CU, is calculated by multiplying E_{tp} by a crop coefficient, K_C, which is dependent on the crop and growth stage, and a soil moisture coefficient, K_S, dependent upon the soil moisture deficiency.

$$CU = K_s \times K_C \times E_{tp} \quad (1)$$

References which give crop and soil moisture coefficients are given at the end of this exercise.

Potential Evapotranspiration

Potential evapotranspiration is determined by measuring the water use of a mature, well-watered reference crop. As was stated, E_{tp} is dependent upon the energy status of the air surrounding the crops. Thus by measuring both E_{tp} and weather parameters which determine this energy status, empirical equations can be developed or theoretical energy balance equations calibrated to predict E_{tp} from weather measurements.

The evapotranspiration equation which utilizes the most weather parameters and is considered to be the most accurate for a wide range of climatic conditions is the modified Penman equation :

$$E_{tp}\ (mm) = .41\ \left[(\Delta/(\Delta + \gamma))\ (R_n + G) + (\gamma\ /\ (\Delta + \gamma))\ 6.43\ W_f\ (e_a - e_d)\right] \quad 2$$

where

Δ = the slope of the vapor pressure-temperature relationship at the given temperature (kPa/°C),

γ = a psychrometer constant (kPa/°C),

R_n = net solar radiation (MJ/m^2) (0.0418 Langlies (cal/cm^2) = 1 MJ/m^2),

G = soil heat flux, positive when away from the soil (MJ/m^2)

W_f = a locally calibrated wind function (dimensionless),

e_a = the average saturated vapor pressure (kPa), and

e_d = the vapor pressure at the mean dew point (kPa).

The inputs for the modified Penman equation are not usually directly measured but are calculated from measured weather parameters of incoming solar radiation, air temperature, relative humidity, and wind run. These relationships are given in Section 12.4.

Weather parameters are interrelated. For example, relative humidity affects both daily temperature fluctuations and solar radiation. Therefore, E_{tp} equations which require fewer parameters are often used, especially when complete weather data are not available.

An E_{tp} equation which requires measurements of only solar radiation and temperature is the Jensen-Haise relationship :

$$E_{tp}\ (mm) = .41 C_T \left(\overline{T} - T_x\right) R_s, \tag{3}$$

where

R_s = incoming solar radiation (MJ/m^2),

$\overline{T}$ = mean daily temperature (°C), and

C_T and T_x are constants calculated from long term climatic records for the location.

The simplest commonly used E_{tp} equation, developed by Blaney and Criddle, assumes the E_{tP} is proportional to the product of the percentage of daylight hours, P (expressed as a decimal) and a factor dependent on the mean air temperature, $\overline{T}\ (^\circ C)$:

$$E_{tp}\ (mm) = C\left[P\ \left(0.45\,\overline{T} + 8\right)\right] \tag{4}$$

where C is a constant dependent on minimum relative humidity, sunshine hours, and daytime wind estimates. Percentage of daylight hours is dependent only on location and date and is available from published tables (Doohrenbos and Pruitt, 1977). The Blaney-Criddle method is normally used only for long-term calculations (monthly) and the constant, C, is determined for classes (low, medium and high) of relative humidity levels, sunshine hours, and daytime wind run.

Since evaporation from a water surface is also dependent on the energy status of the atmosphere, pan evaporation data can given an indication of crop evapotranspiration. However, the pan data must be calibrated for specific locations (climatic conditions). Evaporation from a class A evaporation pan usually varies between 20 and 30 percent greater than E_{tp}.

References listed at the end of the exercise describe in detail how to use these methods for calculating E_{tp}.

Effective Precipitation

Effective precipitation is that rain, snow, or dew deposited on the soil and plant surface and available to meet the evapotranspiration requirement of the plant. Total precipitation, as measured in rain or snow gauges, must be adjusted for runoff and deep percolation, which will depend upon the precipitation amount and intensity, soil intake rates, topography and percedent soil moisture deficiency. Although a portion of the total precipitation will evaporate directly from the soil and plant surfaces, it is still effective to the extent that the evaporation from wet surfaces absorbs energy that would otherwise cause transpiration.

Instrumentation

Weather measurement instrumentation required to determine potential evapotranspiration and carry out irrigation scheduling will depend upon the E_{tp} equation used. The modified Penman method requires incoming solar radiation, relative humidity, temperature, and wind run; the Jensen-Haise equation requires solar radiation and temperature; the Blaney-Criddle method requires measurement of only temperature and long term estimates of relative humidity, sunshine hours, and wind run; and the evaporation pan method requires only pan evaporation. The method chosen will depend upon the accuracy required, the time basis for the estimate (i.e., daily, weekly, or monthly) and the availability and dependability of historical weather data.

Several types of instruments are available to measure the various parameters. Factors which affect the selection of instrumentation include the desired accuracy, the accessibility of the site, the availability and skill of labour to take readings and service the equipment and the purchase price of the equipment.

Temperature

Air temperature can be measured with a simple mercury or alcohol thermometer, a thermoelectric couple or a bimetallic recording thermograph. Maximum-minimum thermometers record the maximum and minimum temperature between servicing visits. The maximum temperature is obtained from a mercury thermometer with a constriction at the neck of the mercury reservoir which allows the mercury to expand into the gauge but not return to the reservoir. The minimum temperature is obtained from an alcohol thermometer containing a marker which will follow the alcohol as it contracts but not as it expands.

A thermometric couple is a junction of two dissimilar metals which products an electric current when heated. The electric current is converted to a voltage and calibrated with temperature. Electronic recording equipment periodically measure the voltage and converts it to a temperature.

A recording thermograph uses a curved metallic strip as the temperature sensing element. One end of the strip is fixed to the instrument base and the other end is attached to a pen arm linkage. Changes in temperature cause expansion or contraction in the strip which is translated into pen movement via the arm linkage. Temperature can be measured over a wide range although it may be necessary to specify a particular sensor for a particular temperature range. Thermograph calibrations should be checked at least monthly with a mercury thermometer.

Relative Humidity

Relative humidity is the quantity of water vapor in the air expressed as a percentage of the amount it could hold at that temperature. Relative humidity can be measured with a sling psychrometer or hygrograph. The sling psychrometer consists of a wet and a dry bulb thermometer which are manually rotated in a vertical plane. The two temperatures are read and a psychrometric slide rule or chart is used to determine the relative humidity. The psychrometer, like the thermometer, gives a reading at one point in time.

The hygrograph makes a continuous record of relative humidity on a chart mounted on a clock driven cylinder. Human hair is the sensing element, the length of which is a function of the relative humidity in the air. The expansion or contraction of the hair is mechanically magnified and transmitted to the chart via a pen and arm linkage. The chart is mounted on a cylinder or drum which is rotated by either a spring wound or an electric clock drive. The chart rotation speed can be selected as 24 hours or 7 days per revolution by changing the gears on the clock assembly. Such a drum chart is shown in Figure 12.1. The calibration of hygrographs should be checked in the field at least every month with a sling psychrometer.

Figure 12.1. Hygrothermograph.

Often a hygrothermograph, which is a combination of a hygrograph and a thermograph, is used to record both temperature and relative humidity as a function of time. Figure 12.1 shows such a device.

Solar Radiation

Total solar radiation is measured with either a pyroheliometer or a pyranometer. The first device, also called a pyranograph, produces a trace on paper of solar raidation over time while the second produces an electric signal which can be converted to solar radiation.

The most common method of measuring solar radiation utilizes the difference in temperature between highly absorptive black surfaces and highly reflective white surfaces. This temperature difference is proportional to the intensity of radiation. In the pyroheliometer, shown in Fig. 12.2, the temperature differential is measured using alternating black and white bimetallic strips which are connected to a pen arm via a mechanical linkage system. The pen makes a trace over time on a rectangular chart mounted on a rotating drum similar to that used in a hygrothermograph. The resulting trace is calibrated as radiation intensity. The total radiation is the integration of the radiation intensity over time, or the area under this trace.

Figure 12.2. Pyroheliometer.

In the star pyranometer, the temperature differential between the black and white surfaces is measured using thermoelectric junctions which produce an electric current proportional to the temperature difference. Electronic integrators convert the signal to units of thermal radiation, sum the readings and display the total on a resettable electromechanical register or digital display.

Another method of measuring solar radiation uses a photo-transistor to generate an electric current proportional to the radiation intensity.

The current is converted to a voltage which is integrated and calibrated to output units of radiation on a printer or display. Figure 12.3 shows a photo-transistor pyranomete.

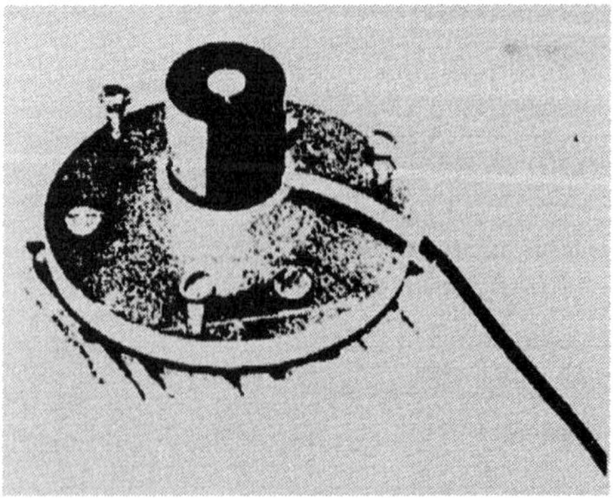

Figure 12.3. Photo transistor pyranometer

Pyroheliometers and pyranometers should be recalibrated every one or two years by comparing the output with the known output of specially calibrated sensors. The United States Bureau of Standards can make such a calibration check.

Due to expense and the difficult of calibrating solar radiation instruments, many areas have historically measured only sunshine hours. Sunshine recorders are transparent spherical bowls which focus sunshine on a specially prepared card. The foucused sunshine burns a band on the card as the sun moves across the sky. The burned band indicates the time and relative intensity of sunshine. Sunshine hours can then be used with the expected clear sky or extraterrestrial radiation for the location to give an estimate of incoming solar radiation.

Wind Run

Wind run, or the distance air moves at a given height over a period of time, is usually measured with a cup anemometer. Cups attached to a vertical shaft rotate in wind. The shaft rotation, after being reduced by a gear assembly, either rotates a mechanical counter or generates a switch closure for each tenth of a kilometer (or tenth of a mile) of air movement which is counted on an electronic register. The anemometer cups are generally mounted 2 meters above the ground. Anemometers can be checked in the field for bearing wear by comparing readings with a second anemometer.

In many areas wind run varies greatly within short distances, so that recorded wind movement may be typical of a relatively limited area.

Evaporation Pan

The standard U.S. Weather Bureau Class A evaporation pan is 121 cm (47 1/2 in.) in diameter and 25.5 cm (10 in.) deep and is installed on a 15 cm (6 in.) wooden platform which allows air to circulate freely around the pan. The pan is constructed of corrosion resistant galvanized metal or stainless steel and is normally filled with water to approximately 6 cm from the top. A stilling well and hook gauge are used to determine the amount of water evaporated. The pan should be refilled (or drained after a rain) to the initial level after each reading.

The water in the pan should be kept relatively clean and free of debris. The pan should not be covered with screen unless appropriate corrections are made when converting the evaporation depths to evapotranspiration. Algae growth can be controlled by periodically adding a small amount of chlorine disinfectant.

Precipitation

Precipitation is measured by collecting rain or snow which falls into a container of a certain diameter. The U.S. Weather Bureau standard nonrecording rain gauge has an 20.3 cm (8 in.) diameter opening. Rain falling into the gauge is funnelled into a 6.4 cm (2.5 in.) diameter measuring tube. This 10.24-to-1 magnification permits reading to the nearest 0.25 mm of precipitation by means of a specially calibrated dip stick which, when inserted into the tube, is wetted to the depth of the water in the tube. After heavy precipitation, overflow from the tube, which is stored in the gauge outer container, can be poured into the emptied tube for measurement. The small exit hole from the funnel decreases evaporation of the collected water, but nonrecording rain gauges should still be measured and emptied daily.

A weighing type rain gauge also uses a 20.3 cm diameter funnel to collect precipitation in a galvanized bucket. A spring scale, which weighs the bucket continuously, is connected to a pen mechanism which records the weight on a chart mounted on a rotating drum. Due to the continuous record, a weighing type rain gauge can be used to determine rainfall intensity as well as total amount. The calibration of weighing rain gauges should be checked every year by placing known weights on the spring scale, measuring the pen travel and making the appropriate adjustments.

A tipping bucket rain gauge uses a 20.3 cm diameter funnel to direct the water into a calibrated bucket. When 0.25 mm of rain has been collected, the bucket tips. During the tip, a mercury switch or magnetic switch is momentarily closed. At the end of the tip, a second bucket is positioned below the funnel, ready to fill and repeat the cycle. a register or electronic counter counts the number of switch closures which is directly proportional to the amount of rainfall. The counter is read daily or following irrigation events, or the switch closures can be accumulated in the memory of electronic recording devices.

Placement of Weather Instruments

The weather instruments are normally grouped together in weather stations. They should be located away from any buildings, trees or other obstructions which could adversely affect the instrument readings. The hygrothermograph and thermometers should be placed in a white, louvered instrument shelter elevated about 1.25 m above the ground surface to allow good air circulation around the instruments while protecting them from direct sunlight and precipitation. The door should open such that direct sunlight does not fall on the thermometers when the door is opened (i.e., open to the north in the northern hemisphere). The rain gauge and anemometer should be level and located away from any obstructions, especially in the direction of the prevailing wind. The pyranometer should be level and securely fastened away from any sunlight obstructions. Pyranometers are often fixed to the roof of instrument shelters. The evaporation pan is generally located in a level grass area away from any surfaces which could greatly increase the advective energy passing over the evaporation pan. Weather stations should generally be fenced to prevent tampering and to keep out animals Figures 12.4 and 12.5 show instruments installed in a weather station.

Temperature, relative humidity and pan evaporation are affected by vegetation in the surrounding area. Consequently, the weather station must be located in a cropped area if crop E_{tp} is being determined, of away from cropped areas if the E_{tp} of natural vegetation is being estimated.

Automated Collection of Weather Data

Recent technological advances in the construction of sensors and data recording equipment make it possible to greatly reduce the labor required to collect weather data. Electronic recording equipment allows frequent automatic collection and recording of instrument readings. The data are converted to the desired engineering units and can be stored in a way to make them directly transferrable to a computer for analysis. The low power requirements of battery-powered recording equipment allow long servicing intervals, making automated collection desirable for sites which are not readily accessible.

Figure 12.4. Instrument shelter with hygrothermograph, maximum and minimum thermometer, and pyroheliometer.

12.2 EQUIPMENT

Note : all instruments should be properly installed in a weather station.

1. pyranometer or pyroheliometer
2. hygrothermograph
3. sling psychrometer and psychromatic table
4. cup anenometer
5. rain gauge
6. class A evaporation pan
7. one week of weather records from the instruments.

Also : thermometer, distilled water

Figure 12.5. Portion of a weather station showing an evaporation pan with point gauge, a weighing rain gauge, and a standard 2 meter and a surface cup anemometer.

12.3 PROCEDURE

Calibration Check

1. Place a thermometer inside the weather shelter out of direct sunlight and away from any obstruction which would interfere with free air movement aroung the thermometer. After equilibrium is reached, compare the thermometer temperature with that indicated on the hygrothermograph and adjust the hydrothermograph accordingly.
2. Saturate the wick on the psychrometer wet bulb thermometer with distilled water. Whirl the psychrometer in a vertical plane at a speed of approximately 30 revolutions per minute until the thermometers have reached equilibrium. Record the relative humidity from either printed tables or a psychromatic slide rule. Compare the value with the relative humidity value on the hygrothermograph. There is a considerable lag time for the hygrothermograph to respond to changes in humidity so make adjustments spar-

ingly. An alternate and more accurate method of calibration is to place the hygrothermograph in a calibration chamber where temperature and humidity can be held constant. Because human hair is fragile and skin oil will affect its calibration, use extreme care and do not touch the hair sensing element when making adjustments.

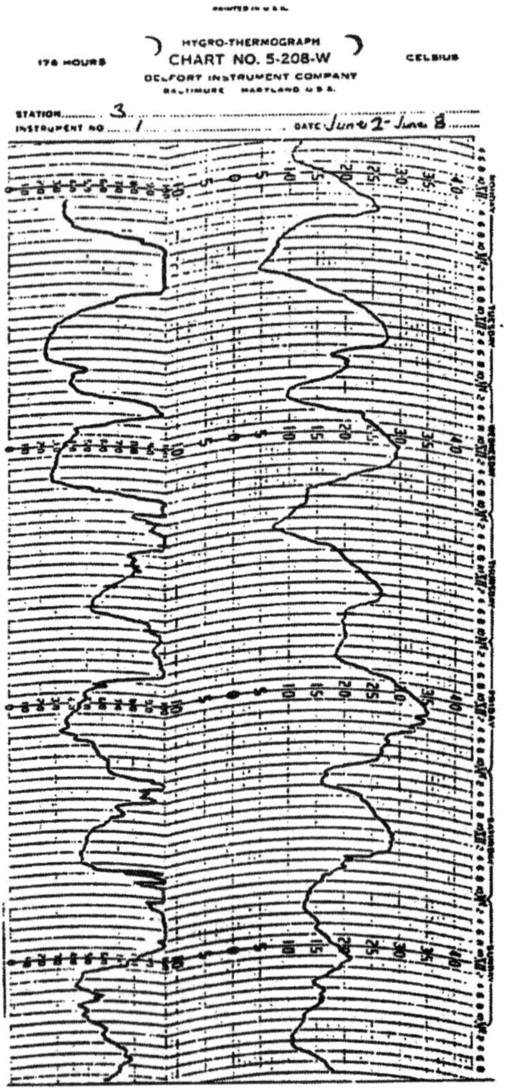

Figure 12.6. Sample hygrothermograph chart showing relative humidity and temperature for one week.

Collecting the Data

3. The maximum and minimum temperatures for each day, are read directly from the hygrothermograph chart. Figure 12.6 shows a typical hygrothermograph chart with temperature on the upper scale and relative humidity on the lower scale. Record the daily maximum and minimum temperatures from the hygrothermograph chart on the data collection sheet provided. Also record the temperatures and relative humidities for 2:00, 6:00, 10:00, 14:00, 18:00 and 22:00 hours. Use Table 12.1 to find the dew point (that temperature at which the vapor pressure reaches the saturation vapor pressure or the relative humidity reaches 100 percent) for each of the six readings. Calculate the average des point temperature for each day.

Table 12.1 Dew point temperature for given temperatures and relative humidities.

DEW POINT TEMERATURES (DEG C)																				
RELATIVE HUMIDITY %																				
TEMP	5	10	15	20	25	30	35	40	45	50	55	60	65	70	75	80	85	90	95	100
0.0	-00.0	-00.0	-00.0	-00.0	-00.0	-15.5	-13.5	-12.0	-10.6	-9.2	-8.0	-6.9	-5.9	-4.9	-4.0	-3.1	-2.3	-1.5	-0.8	0.0
1.0	-00.0	-00.0	-00.0	-00.0	-00.0	-14.5	-12.6	-11.2	-9.7	-8.4	-7.1	-6.0	-4.9	-3.9	-3.0	-2.2	-1.3	-0.5	0.2	1.0
2.0	-00.0	-00.0	-00.0	-00.0	-15.9	-13.7	-11.9	-10.2	-8.7	-7.4	-6.2	-5.0	-3.9	-2.9	-2.1	-1.2	-0.3	0.5	1.2	2.0
3.0	-00.0	-00.0	-00.0	-00.0	-15.0	-12.8	-11.0	-9.3	-7.9	-6.5	-5.2	-6.1	-3.0	-2.0	-1.1	-0.2	0.6	1.4	2.2	3.0
4.0	-00.0	-00.0	-00.0	-00.0	-14.2	-12.0	-10.1	-8.5	-7.0	-5.6	-4.3	-3.1	-2.1	-1.1	-0.1	0.8	1.6	2.5	3.2	4.0
5.0	-00.0	-00.0	-00.0	-16.1	-13.4	-11.2	-9.2	-7.5	-6.0	-4.6	-3.3	-2.2	-1.2	-0.1	0.8	1.8	2.6	3.4	4.1	5.0
6.0	-00.0	-00.0	-00.0	-15.2	-12.5	-10.4	-8.4	-6.6	-5.1	-3.7	-2.5	-1.2	-0.2	0.9	1.9	2.7	3.5	4.4	5.1	6.0
7.0	-00.0	-00.0	-00.0	-14.3	-11.7	-9.5	-7.5	-5.8	-4.2	-2.7	-1.5	-0.3	0.8	1.9	2.8	3.7	4.6	5.4	6.2	7.0
8.0	-00.0	-00.0	-00.0	-13.5	-10.9	-8.6	-6.6	-4.8	-3.3	-1.9	-0.6	0.6	1.8	2.8	3.7	4.7	5.6	6.5	7.3	8.0
9.0	-00.0	-00.0	-16.2	-12.7	-10.1	-7.8	-5.7	-3.9	-2.4	-1.0	0.4	1.6	2.7	3.7	4.7	5.6	6.7	7.6	8.3	9.0
10.0	-00.0	-00.0	-15.4	-12.0	-9.2	-6.9	-4.8	-3.0	-1.4	0.0	1.3	2.5	3.6	4.7	5.7	6.8	7.7	8.5	9.4	10.0
11.0	-00.0	-00.0	-14.5	-11.2	-8.4	-6.0	-3.9	-2.2	-0.5	0.9	2.2	3.5	4.5	5.6	6.8	7.7	8.7	9.5	10.3	11.0
12.0	-00.0	-00.0	-13.8	-10.4	-7.5	-5.1	-3.0	-1.2	0.4	1.9	3.1	4.4	5.6	6.8	7.8	8.8	9.6	10.5	11.3	12.0
13.0	-00.0	-00.0	-12.9	-9.5	-6.6	-4.2	-2.2	-0.3	1.2	2.7	4.0	5.3	6.6	7.8	8.8	9.7	10.7	11.5	12.3	13.0
14.0	-00.0	-00.0	-12.3	-8.7	-5.8	-3.3	-1.3	0.5	2.2	3.6	5.1	6.4	7.6	8.7	9.7	10.7	11.7	12.5	13.3	14.0
15.0	-00.0	-16.3	-11.5	-7.9	-4.9	-2.5	-0.4	1.4	3.0	4.6	6.0	7.4	8.6	9.7	10.7	11.6	12.6	13.5	14.3	15.0
16.0	-00.0	-15.5	-10.7	-7.0	-4.1	-1.6	0.5	2.4	4.0	5.5	7.1	8.3	9.5	10.7	11.6	12.7	13.6	14.5	15.3	16.0
17.0	-00.0	-14.8	-9.9	-6.2	-3.2	-0.8	1.3	3.2	4.9	6.5	8.0	9.3	10.5	11.6	12.7	13.6	14.6	15.5	16.3	17.0
18.0	-00.0	-14.0	-9.1	-5.4	-2.4	0.1	2.2	4.1	5.8	7.5	8.9	10.2	11.5	12.6	13.6	14.6	15.6	16.4	17.3	18.0
19.0	-00.0	-13.2	-8.3	-4.5	-1.5	1.0	3.1	5.0	6.9	8.4	9.9	11.2	12.4	13.5	14.6	15.7	16.5	17.4	18.3	19.0
20.0	-00.0	-12.5	-7.4	-3.7	-0.7	1.9	4.0	5.9	7.8	9.4	10.8	12.1	13.3	14.5	15.5	16.5	17.5	18.4	19.3	20.0
21.0	-00.0	-11.9	-6.6	-2.8	0.2	2.7	4.9	7.0	8.7	10.1	11.7	13.0	14.1	15.4	16.5	17.5	18.5	19.4	20.3	21.0

22.0	–00.0	–11.1	–5.9	–2.1	1.0	3.5	5.7	7.9	9.6	11.2	12.7	14.0	15.2	16.4	17.4	18.5	19.5	20.4	21.3	22.0
23.0	–00.0	–10.4	–5.0	–1.2	1.9	4.4	6.8	8.8	10.5	12.1	13.6	14.9	16.2	17.3	18.4	19.5	20.5	21.4	22.2	23.0
24.0	–00.0	–9.6	–4.3	–0.4	2.7	5.3	7.7	9.7	11.5	13.0	14.5	15.9	17.1	18.3	19.4	20.5	21.4	22.3	23.1	24.0
25.0	–00.0	–8.8	–3.5	0.4	3.6	6.2	8.4	10.6	12.4	14.0	15.5	16.8	18.1	19.3	20.4	21.4	22.3	23.2	24.1	25.0
26.0	–00.0	–8.0	–2.4	1.2	4.4	7.2	9.5	11.5	13.2	14.9	16.3	17.8	19.0	20.2	21.4	22.3	23.2	24.1	25.0	26.0
27.0	–15.7	–7.3	–1.9	2.0	5.2	8.1	10.3	12.4	14.2	15.8	17.3	18.7	20.0	21.2	22.2	23.2	24.2	25.1	26.0	27.0
28.0	–15.0	–6.5	–1.1	2.9	6.1	8.9	11.2	13.2	15.1	16.8	18.2	19.6	20.9	22.1	23.1	24.1	25.2	26.1	27.0	28.0
29.0	–14.3	–5.8	–0.3	3.7	7.1	9.8	12.1	14.2	16.0	17.4	19.2	20.5	21.9	22.8	24.0	25.0	26.1	27.1	28.0	29.0
30.0	–13.7	–5.0	0.5	4.6	7.9	10.7	12.9	15.1	16.9	18.6	20.1	21.5	22.6	23.8	24.9	26.1	27.1	28.0	29.0	30.0
31.0	–12.9	–4.2	1.2	5.3	8.8	11.5	13.9	15.9	17.8	19.5	21.0	22.3	23.6	24.8	25.9	27.0	28.0	29.0	29.9	31.0
32.0	–12.4	–3.5	2.0	6.2	9.6	12.4	14.8	16.9	18.7	20.4	21.9	23.2	24.4	25.7	26.9	28.0	29.0	30.0	31.0	32.0
33.0	–11.7	–2.7	2.9	7.2	10.5	13.2	15.6	17.8	19.6	21.3	22.7	24.1	25.4	26.7	27.8	28.9	30.0	31.0	32.0	33.0
34.0	–11.0	–2.0	3.6	8.0	11.3	14.1	16.4	18.6	20.5	22.2	23.6	25.0	26.1	27.6	28.8	29.9	31.0	32.0	32.9	34.0
35.0	–10.3	–1.3	4.5	8.8	12.1	15.0	17.3	19.5	21.4	22.9	24.4	26.0	27.3	28.5	29.8	30.9	32.0	32.9	34.2	35.0
36.0	–9.6	–0.4	5.2	9.6	12.9	15.8	18.2	20.4	22.2	23.8	25.4	26.9	28.2	29.5	30.7	31.9	32.9	39.2	35.2	36.0
37.0	–8.9	0.3	56.0	10.4	13.8	16.7	19.1	21.3	22.9	24.7	26.3	27.8	29.2	30.5	31.6	32.9	34.1	35.3	36.2	37.0
38.0	–8.2	1.1	7.0	11.1	14.7	17.5	20.0	22.1	23.8	25.6	27.2	28.7	30.1	31.4	32.6	34.0	35.2	36.2	37.2	38.0
39.0	–7.5	1.8	7.8	12.1	15.5	18.1	20.9	22.8	24.8	26.5	28.1	29.7	31.1	32.4	33.8	35.0	36.1	37.2	38.2	39.0
40.0	–6.9	2.5	8.6	12.8	16.3	19.3	21.8	23.7	25.6	27.4	29.1	30.7	32.0	33.4	34.9	36.0	37.1	38.2	39.2	40.0
41.0	–6.1	3.3	9.6	13.7	17.2	20.1	22.5	24.6	26.6	28.3	30.0	31.6	34.9	34.5	35.8	37.0	38.1	39.2	40.1	41.0
42.0	–5.5	4.1	10.1	14.5	18.0	21.0	24.2	25.5	27.5	29.1	30.9	32.5	35.5	35.5	36.7	37.9	39.0	40.1	41.2	42.0
43.0	–5.7	4.8	14.9	15.3	18.8	21.8	25.1	26.3	28.3	30.2	31.9	33.5	36.5	36.5	37.7	38.9	40.0	41.1	41.9	43.0
44.0	– 6.0	5.4	15.7	16.1	19.6	22.5	25.0	27.2	29.3	31.1	32.8	34.6	37.4	37.4	38.7	39.9	41.0	41.9	42.8	44.0

4. The total solar radiation for each day is calculated b determining the area under a radiation intensity reace from the pyroheliometer. Figure 12.7 shows a typical pyroheliometer chart with the total solar radiation calculated for one day. The radiation intensity is the chart value multiplied by the instrument constant, K. The instrument constant is supplied by the manufacturer or is determined when the instrument is recalibrated. The total energy density is the radiation intensity multiplied by the time duration and is represented by the area under the curve. The area can be determined by numerical integration, or can be planimetered and converted to energy density units by multiplying the area by a conversion factor. The conversion factor is determined by calculating the energy density of a known area and multiplying by the ratio of the actual area to the known area (cross-hatched area in Figure 12.7). Calculate the appropriate conversion factor, paying particular attention to the units involved, for the chart given to you. Planimeter the area beneath the

curve for each day and calculate the total solar radiation for each day.

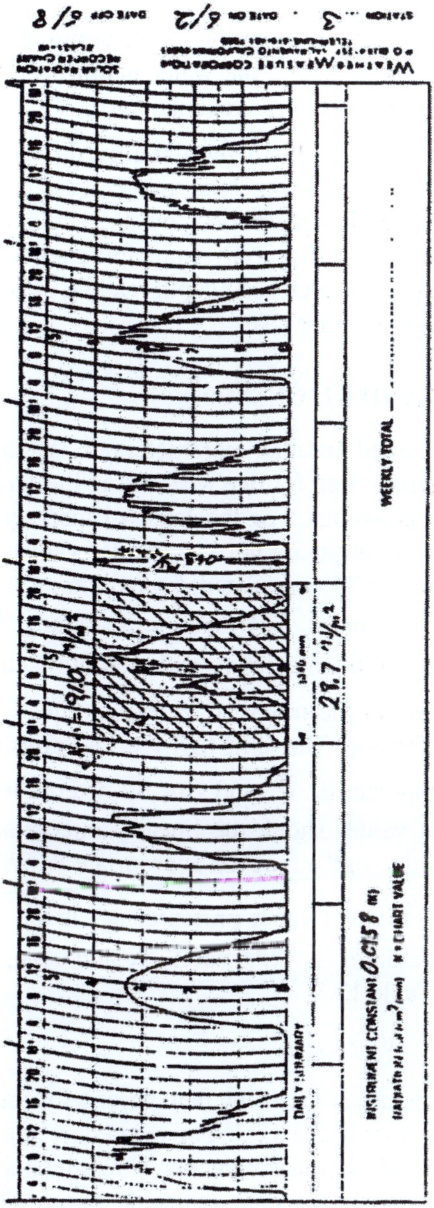

Figure 12.7. Sample pyroheliometer chart showing incoming solar radiation intensity for one week.

5. The daily wind run is calculated by subtracting the previous day's reading from the current day's reading. Readings should be taken at the same time each time.
6. The daily pan evaporation is determined by measuring, with the hook gauge, the water depth decrease in the pan since the previous day's reading.
7. Daily precipitation is determined by direct reading with a dipstick (nonrecording type raingauge), or the increased counter reading multiplied by 0.25 mm/count (tipping bucket type) after any precipitation event, or by determining the increased depth reading on a weighing type rain gauge recorder chart.

12.4 ANALYSIS AND DISCUSSION

Use the measured weather parameters to calculate the daily potential evapotranspiration for one week using the Jensen-Haise and modified Penman equations. For the Jensen-Haise equation (Eq. 3), if constants for the location are not known, use values of $C_T = 0.024$ mm/°C and $T_x = -8.9$°C (calculated from long-term climatic data from Ft. Collins, Colorado). The mean temperature, $\overline{T}\ (^\circ C)$, can be taken as the average of the maximum and minimum temperature.

The parameters in the modified Penmann equation (Eq. 2) can be determined as follows: The term $(\Delta/(\Delta + \gamma))$ varies with elevation and the mean temperature, $\overline{T}$, and can be determined from Table 12.2. Intermediate values should be linearly interpolated. The term $(\gamma/(\gamma + \Delta))$ is equal to one minus the appropriate $(\Delta/(\Delta + \gamma))$ value.

The average saturated vapor pressure, e_a, is the average of the saturated vapor pressures, e_s (kPA) at T_{MAX} and T_{MIN}, which can be determined from Table 12.3.

$$e_a = \left[e_{s-T_{MAX}} + e_{s-T_{MIN}}\right]/2 \tag{5}$$

The vapor pressure at the mean dew point temperature, e_d (kPa), is also determined from Table 12.3. The mean dew point temperature is calculated from 6 daily measurements of temperature and relative humidity as indicate by the data collected sheet. The dew point temperatures are determined from the temperature and relative humidity using Table 12.1.

The net solar radiation, R_n, can be calculated as :

$$R_n = 0.77\ R_s - \left[\frac{0.9R_s}{R_{so}} + 0.1\right] R_{bo} \tag{6}$$

where : R_s = the measured incoming solar radiation (MJ/m^2),

$$R_{SO} = 30.1 \exp\left[-\left(\frac{D-170}{B}\right)^2\right] \left(\text{expected clear sky radiation } \left(MJ/m^2\right)\right),$$

B = 279 – 2.98 × Latitude (°)

D =day of the year (Julian day)

$$R_{bo} = 2.45 \times 10^{-9}\left[0.37 - 0.139\sqrt{e_d}\right]\left(T_{MAX} + 273.15\right)^4 + \left(T_{MIN} + 273.15\right)^4$$

(net longwave radiation, MJ/m^2) and

e_d = the vapor pressure at the mean dew point temperature (kPa).

Table 12.2. Variation of $\Delta / (\Delta + \gamma)$ with elevation and temperature (from ASAE, 1980).

			Elev.,			
°C	0	500	1000	1500	2000	2500
0.0	0.401	0.414	0.428	0.443	0.458	0.475
5.0	0.471	0.491	0.505	0.520	0.536	0.552
10.0	0.551	0.564	0.578	0.593	0.608	0.624
15.0	0.620	0.632	0.645	0.659	0.673	0.688
20.0	0.681	0.693	0.705	0.717	0.730	0.743
25.0	0.735	0.745	0.756	0.767	0.778	0.790
30.0	0.781	0.790	0.799	0.809	0.818	0.828
35.0	0.820	0.828	0.835	0.844	0.852	0.860
40.0	0.852	0.858	0.867	0.872	0.879	0.886
45.0	0.878	0.884	0.889	0.895	0.901	0.907
50.0	0.900	0.904	0.909	0.914	0.919	0.924

$$\frac{\gamma}{\Delta + y} = 1 - \frac{\Delta}{\Delta + \gamma}, \text{ based on the U.S. standard atmosphere.}$$

Table 12.3. Saturation vapor pressure over water at various temperatures.

SATURATION VAPOR PRESSURE OVER WATER Pressures in kiloPascals (kPa)					
Temp. deg. C	.0	.2	Tenths .4	.6	.8
0.0	0.61	0.62	0.62	0.62	0.63
1.0	0.66	0.66	0.66	0.67	0.68
2.0	0.71	0.71	0.71	0.72	0.73
3.0	0.76	0.76	0.76	0.77	0.78
4.0	0.81	0.82	0.82	0.83	0.84
5.0	0.87	0.88	0.88	0.89	0.90
6.0	0.93	0.94	0.94	0.95	0.96
7.0	1.00	1.01	1.01	1.02	1.03
8.0	1.07	1.08	1.08	1.09	1.10
9.0	1.15	1.16	1.16	1.17	1.18
10.0	1.23	1.24	1.24	1.25	1.26
11.0	1.31	1.32	1.33	1.34	1.35
12.0	1.40	141	1.42	1.43	1.44
13.0	1.50	1.51	1.52	1.53	1.54
14.0	1.60	1.61	1.62	1.63	1.64
15.0	1.70	1.72	1.73	1.74	1.75
16.0	1.82	1.83	1.84	1.85	1.86
17.0	1.94	1.95	1.96	1.97	1.99
18.0	2.05	2.08	2.09	2.10	2.12
19.0	2.20	2.21	2.22	2.24	2.25
20.0	2.34	2.35	2.37	2.38	2.40
21.0	2.49	2.50	2.52	2.53	2.55
22.0	2.64	2.66	2.68	2.69	2.71
23.0	2.81	2.83	2.84	2.86	2.88
24.0	2.95	3.00	3.02	3.04	3.06
25.0	3.17	3.19	3.21	3.22	3.24
26.0	3.36	3.38	3.40	3.42	3.44
27.0	3.56	3.59	3.61	3.63	3.65
28.0	3.78	3.80	3.82	3.85	3.87
29.0	4.01	4.03	4.05	4.08	4.10
30.0	4.24	4.27	4.29	4.32	4.34
31.0	4.49	4.52	4.54	4.57	4.60
32.0	4.76	4.78	4.81	4.84	4.85
33.0	5.03	5.06	5.09	5.12	5.14
34.0	5.32	5.35	5.38	5.41	5.44
35.0	5.62	5.65	5.69	5.72	5.75
36.0	5.94	5.97	6.01	6.08	6.07

37.0	6.28	6.31	6.34	6.38	6.41
38.0	6.63	6.66	6.70	6.73	6.77
39.0	6.99	7.03	7.07	7.11	7.14
40.0	7.38	7.42	7.46	7.50	7.54
41.0	7.78	7.82	7.86	7.90	7.95
42.0	8.20	8.24	8.28	8.33	8.38
43.0	8.44	8.69	8.73	8.78	8.82
44.0	9.10	9.15	9.20	9.25	9.29
45.0	9.59	9.63	9.68	9.73	9.76

The daily soil heat flux, G, can be estimated from the difference between the mean of the mean daily temperature of the previous three days $\left(\overline{T}_{-1} + \overline{T}_{-2} + \overline{T}_{-3}\right)$ and the present daily mean temperature, $\overline{T}$:

$$G = 0.377\left[\left(\overline{T}_{-1} + \overline{T}_{-2} + \overline{T}_{-3}\right)/3 - \overline{T}\right] \tag{7}$$

If the mean temperature of the previous three days is not known, assume they equal the weekly mean for the available data, or assume G = 0. The locally calibrated wind function is a linear function of the measured daily wind run, W(km). For Fort Collins the function is :

$$W_f = 1.1 + 0.0106\, W \tag{8}$$

Compare the E_{tp} values given by the two methods. Is one consistently higher or lower than the other? If a particular day were especially windy, which method would give the higher E_{tp} value? Which would be more accurate? Why?

Assume a crop coefficient, K_C, for a mature corn crop of 0.95 and a soil coefficient for the well watered field to be 1.0. What is the crop water use for the week? If the soil field capacity is 25 percent, the wilting point is 12%, the bulk density is 1.3, the root zone is 1 meter, and the acceptable depletion level is 50 percent, at the present crop water use rate, how many can elapse between irrigations?

If you were scheduling irrigations for a 500 ha irrigated area, would you use weather measurements or soil water measurements?

Potential Evapotranspiration Calculation Example

Calculate the E_{tp} for the fourth day of the week, June 5, shown on Figures 12.6 and 12.7. The anemometer readings for the week are shown on the example data sheet (Fig. 12.8). Since wind run is measured at 12:00, the wind run for the day would be the average of the wind run for the present and next day's readings, or the next day's reading minus the past day's reading divided by 2.

$$W = \frac{3221 - 3015}{2} = 103\ km$$

The temperature and relative humidity at 2 : 00, 6:00, 10:00, 14:00, 18:00 and 22:00 hrs is read from the hygrothermograph chart (Fig. 12.6) and a dew point is determined for each set of readings from Table 12.1. The average dewpoint is calculated as 13.9°C and recorded on the sample data sheet. The maximum and minimum temperature is also read from the chart and recorded on the data sheet. The two temperatures are averaged to give the mean temperature

$$\overline{T} = \frac{T_{MAX} + T_{MIN}}{2} = \frac{27^\circ C + 7^\circ C}{2} = 17^\circ C$$

Jensen-Haise

$$E_{tp} = .41C_T \left[\overline{T} - T_x\right] R_S = .41\,(.024)\left[17 - (-8.9)\right]27.8 = 7.1\ mm$$

Modified Penman

At an elevation of 1500 m and for $\overline{T} = 17^\circ C$ (Table 12.2) :

$$\Delta/(\Delta + \gamma) = 0.68$$

$$\gamma\,/\,(\gamma + \Delta) = 1 - \Delta\,/\,(\Delta + \gamma) = 1 - 0.68 = 0.32$$

From Table 12.3 :

$$e_{s-T_{MAX}} = e_{s_{27^\circ C}} = 3.56\ kPa$$

$$e_{s-T_{MIN}} = e_{s7^\circ C} = 1.00\ kPa,$$

so : $e_s = \left[e_{s-T_{MAX}} + e_{s-T_{MIN}}\right]/2 = \left[3.56 + 1.00\right]/2 = 2.28\ kPa$

Since June 5 is the 156th day of the year and Ft. Collins is at 40.2°N. Latitude :

$$R_{so} = (30.1)\exp\left[-\left(\frac{156 - 170}{159}\right)^2\right] = 29.9\ MJ/m^2.$$

Soil-Water Engineering Laboratory
Exercise 12
Weather Measurement Data Sheet

Name ____________________

Location ____________________

Time \ Date								
0200	T							
	RH							
0600	T							
	RH							
1000	T							
	RH							
1400	T							
	RH							
1800	T							
	RH							
2200	T							
	RH							
Average Dew Point Temperature °C								
Maximum Temperature °C								
Minimum Temperature °C								
Mean Temperature °C								
Solar Radiation	Area							
	$\frac{MJ}{m^2}$							
Anemometer Rdg.								
Wind run, W-km								
Pan Evaporation (mm)								
Precipitation (mm)								
E_{tp} (mm) Jensen-Haise								
Modified Penman								

Figure 12.8. Data sheet for calculation example.

From Table 12.3, the saturation vapor pressure at the average dew point temperature (13.9°C) = 1.59 kPa, so :

$$R_{bo} = 2.45 \times 10^{-9} \left[0.37 - 0.139 \sqrt{e_d}\right] \left[(T_{MAX} + 273.15)^4 + (T_{MIN} + 273.15)^4\right]$$

$$= 2.45 \times 10^{-9} \left[0.37 - 0.139 \sqrt{1.59}\right] \left[(27 + 273.15)^4 + (7 + 273.15)^4\right]$$

$$= 6.8 MJ / m^2,$$

and :

$$R_n = 0.77 \, R_s - \left[\frac{0.9 \, R}{R_{so}} + 0.1\right] R_{bo} = 0.77 \times 27.8 - \left[\frac{0.9 \times 27.8}{29.9} + 0.1\right] 6.8$$

$$= 15.0 \; MJ / m^2.$$

The average temperature for the previous three days is calculated from daily maximum and minimum temperature as shown on the data sheet so :

$$G = 0.377 \left[(\overline{T}_{-1} + \overline{T}_{-2} + \overline{T}_{-3})/3 - \overline{T}\right] = 0.377 \left[(18 + 17 + 20)/3\right] - 17$$

$$= 0.50 \; MJ / m^2$$

From Eq. 8 and the calculated wind run of 103 km

$$W_f = 1.1 + 0.0106 \; W = 1.1 + 0.0106 \; (103) = 2.19,$$

so :

$$E_{tp} = 0.41 \left[(\Delta / (\Delta + \gamma)) (R_n + G) + (\gamma / (\gamma + \Delta)) \; 6.43 \; W_f \; (e_a - e_d)\right]$$

$$= 0.41 \left[0.68 \; (15.0 + 0.50) + 0.32 \times 6.43 \times 2.19 \; (2.28 - 159)\right]$$

$$= 5.6 \; mm.$$

12.5 REFERENCES

GENERAL

American Society of Agricultural Engineers (ASAE) 1980. Design and Operation of Farm Systems. M.E. Jensen, ed. Chapter 16, ``Water Required" by R.D. Burman, P.R. Nixon, J.L. Wright, and W. Witt. ASAE, P.O. Box 410, St. Joseph, Mi. 49085.

American Society of Civil E. (ASCE). 1973. Consumptive Use of Water and Irrigation Requirements. M.E. Jensen, ed. ASCE, 345 East 47th St. , New York 10017.

Doohrenbos, J. and W.O. Pruitt, 1977. Crop Water Requirements. FAO Irrigation and Drainage Paper #24. FAO, Rome.

Hargreaves, G.H. 1977. World Water for Agriculture. Prepared for the U.S. Agency for Intern'l Development at Utah State University, Logan, Utah.

World Meteorological Organization (WMO). 1970. Guide to Hydrometeorological Practices. WMO#168, TP. 82, 2nd ed., Geneva, Switzerland.

Instrumentation

World Meteorological Organization (WMO). 1971. Guide to Meteorological Instrument and Observing Practices. WMO#8, TP.3, 4th ed. Geneva, Switzerland.

Crop and Soil Coefficients

ASAE (1980)

ASCE (1973)

Doorenbos and Pruitt (1977)

Kincaid, D.C. and D.F. Heerman. 1974. Scheduling Irrigations using a Programmable Calculator. ARS-NC-12. USDA/ARS.

Irrigation Scheduling

American Society of Agricultural Engineers (ASAE). 1981. Irrigation Scheduling for Water and Energy Conservation in the 1980s. Proceedings of the Irrigation Scheduling Conference. December 14.15, 1981, Palmar House, Chicago, Illinois.

California, Dept. of Water Resources. 1980. Crop Water Use, A Guide for Scheduling Irrigtion in the Southern San Joaquin Valley, 1977-1979. Calif. Dept. of Water Resources, Sacramento, Calif.

Kincaid and Heermann (1974)

Calculating Potential Evapotransporation

ASAE (1980)

Doohrenbos and Pruitt (1977)

ASCE (1973)

APPENDIX 1

REPEATEDLY USED LABORATORY EQUIPMENT

EQUIPMENT	EXERCISE NO. IN WHICH EQUIPMENT IS USED											
	1	2	3	4	5	6	7	8	9	10	11	12
Soil sampling cans	x	x	x	x	x	x						
Oven	x	x	x	x	x	x				x		
Balance $(\pm 0.1\, gm)$	x	x	x	x	x	x						
Dessicator	x	x	x	x	x	x				x		
Spade or shovel	x								x		x	
Meter stick	x						x				x	
Laboratory Scoop			x							x		
Spatula	x			x								
Mortar and Pestle		x	x	x								
#6 sieve		x		x						x		
Thermometer		x	x							x		x
Wash bottle		x	x									
250 ml beakers		x	x									
Distilled or deionized water			x			x						x
De-aerated water		x				x						
Watch			x						x			
Stop watch							x	x			x	
Vacuum system		x								x		

APPENDIX 2

STATISTICAL ANALYSIS OF MEASUREMENTS

Introduction

The purpose of any soil sampling is to obtain information about a particular soil. The sample usually is only a small portion of the entire soil mass in which you are interested. This large mass of soil is called the ``population" or ``universe". Information from the sample is of interest only in the sense that it yields information about the population.

There are certain characteristics in each population which describe it. The true value of each such characteristic in the population is called a parameter. The purpose of sampling is to estimate these parameters.

Measures of Central Tendency

Several measures have been developed to represent central tendency, but only two the arithmetic average and the median-will be explained here in detail. When the data are not skewed, these two measures will be about equal numerically , and either is a good indicator of central tendency. However, when data are skewed, the median may be a better measure than the arithmetic mean.

The Mean or Arithmetic Average

The mean or arithmetic average is defined as the sum of all measurements divided by the number of measurements,

$$\overline{X} = \sum_{i=1}^{n} x_1 / n, \tag{A-1}$$

where $\overline{X}$ is the symbol for the sample mean, x_1 is the ith measurement and n is the total number of measurements.

The Median

The median is used as a measure of central tendency when the distribution of measurements is skewed (i.e., a few observations are quite large or small in comparison to the others).

(a) If n is odd the median is calculated by ordering the measurements from minimum to maximum, and then selecting the (n + 1)/2 measurement. As an example, suppose we have the following measurements.

before ordering : 5, 6, 5, 4, 8, 9, 3

after ordering : 3, 4, 5, 5, 6, 8, 9

the median is equal to 5.

(b) If n is even the median is calculated by ordering the measurements from minimum to maximum, and then averaging the n/1 and n/2 + 1 measurements. As an example, suppose we have the following measurements.

before ordering : 7, 6, 8, 5, 3, 10

after ordering : 3, 5, 6, 7, 8, 10

the median is equal to (6 + 7)/2 = 6.5

Measures of Variability : Spread or Dispersion

The Variance : S^2

The variance is the average of the squares of the deviations measurements about their mean. The definition formula for variance is

$$S^2 = \frac{\sum_{i=1}^{n} \left(x_i - \overline{X}\right)^2}{n - 1} \qquad \text{(A-2)}$$

Note that the average is found by dividing by n-1, not n.

The Standard Deviation, S

The standard deviation is the positive square root of the variance and is related to how far, on the average, a typical observation lie from the sample mean.

$$S = \sqrt{S^2} \tag{A-3}$$

Statistical Inference

The objective of statistics is to make inference (prediction of decision) about the population parameters based upon informatic collected in the sample. We may estimate or predict the value of the parameter using an estimator. The estimator is a rule, general expressed as a formula, that tells us how to calculate the estimate based upon information in the sample.

Estimation procedures may be divided into two types, the point estimate and the interval estimate. The point estimate uses the information in the sample to arrive at a single number or point. The interval estimate uses the information in the sample to calculate two points which are to enclose the true value of the parameter estimated.

Point Estimators

The best estimate of the population mean, μ, is the sample mean, $\overline{X}$ (Eq. A-1), the best estimate of the population variance, σ^2, is the sample variance, S^2 (Eq. A-2). The best estimate of the population standard deviation, σ, is the sample standard deviation, S (Eq. A-3).

Interval Estimation for the Population Mean

The interval estimation or confidence interval for a population mean may be easily obtained from the point estimates described above. If n is small (n less than 30) the confidence interval for the population mean μ, is calculated using the student's t-test.

$$\overline{X} \pm t_\alpha \, S / \sqrt{n}, \tag{A-4}$$

where t_α is the test statistic listed in Table A-1 and α is the degree of confidence desired. This implies that there is α percent probability that the true mean, μ, lies within this confidence interval calculated from the data sample.

Table A-1. Common tests statistic values.

n	t_{90}	t_{95}	t_{99}
2	6.314	12.706	63.657
3	2.920	4.303	9.925
4	2.353	3.182	5.841
5	2.132	2.776	4.604
6	2.015	2.571	4.032
7	1.943	2.447	3.707
8	1.895	2.365	3.499
9	1.860	2.306	3.355
10	1.833	2.262	3.250

Method of Least Squares (Regression Analysis)

One of the most common methods to fit functional relationships to experimental data is the method of least squares. In this type of fitting, the sum of the squares of the deviations of observed values from some functional form is minimized.

The experimental observations consists of a set of n data points (X_i, Y_i) in which Y_i is an observed value corresponding to an independent variable Xi.

Application of Least Squares to a Straight Line (Linear Regression)

If a straight line is desired to fit a set of n data points (X_i, Y_i), the functional form is,

$$Y = a + mX \tag{A-5}$$

where Y is the dependent variable, X is the independent variable and a and m are constants defined by :

$$m = \frac{n \sum_{i=1}^{n} X_i Y_i - \sum_{i=1}^{n} X_i \sum_{i=1}^{n} Y_i}{n \sum_{i=1}^{n} X_i^2 - \left(\sum_{i=1}^{n} X_i \right)^2} \tag{A-6}$$

and

$$a = \frac{\sum_{i=1}^{n} Y_i \sum_{i=1}^{n} X_i^2 - \sum_{i=1}^{n} X_i Y_i \sum_{i=1}^{n} X_i}{n \sum_{i=1}^{n} X_i^2 - \left(\sum_{i=1}^{n} X_i \right)^2} = \frac{\sum_{i=1}^{n} Y_i - m \sum_{i=1}^{n} X_i}{n}$$

(A-7)

The coefficient of determination, r^2, indicates how closely the equation fits the experimental data.

$$r^2 = \frac{\left[n \sum_{i=1}^{n} X_i Y_i - \sum_{i=1}^{n} X_i \sum_{i=1}^{n} Y_i \right]^2}{\left[n \sum_{i=1}^{n} X_i^2 - \left(\sum_{i=1}^{n} X_i \right)^2 \right] \left[n \sum_{i=1}^{n} Y_i^2 - \left(\sum_{i=1}^{n} \right) Y_i \right]^2} \quad \text{(A-8)}$$

the value of r^2 will lie between 0 and 1, the closer r^2 is to 1, the better the fit.

In the event that a = 0, the line has the equation

$$Y = mX,$$

and

$$m = \sum_{i=1}^{n} X_i Y_i \,/ \sum_{i=1}^{n} X_i^2 \quad \text{(A-9)}$$

Linear Regression of Exponential and Power Curve Relationships

Nonlinear data can often the ``linearlized" so that linear regression techniques can be used to determine best fit relationships. This is commonly done graphically when data is plotted on log-log or semilog graph paper to determine a straight line relationship. For example, if data fits a power curve of the form :

$$Y = a_1 \, X^{b_1}, \quad \text{(A-10)}$$

the data can be linearlized by taking the logrithms of both sides :

$$\log Y = \log a_1 + b_1 \log X \quad \text{(A-11)}$$

Such data would plot as a straight line on log-log paper. Similarly, data which fits an exponential curve :

$$Y = a_2\ e^{b_2 X}, \tag{A-12}$$

can be linearlized by taking the logs of both sides.

$$\log Y = \log a_2 + b_2\ X \tag{A-13}$$

Such a relationship plots linearly on semilog paper.

In order to generate the coefficients of Eq. A-10 using linear regression techniques, take the logs of both X_i and Y_i before inserting into the regression equations (Eq. A-6 to A-8). As indicated by Eq. A-11, a_1, will be the antilog of the derived intercept, a, and the exponent, b_1, will be the linear slope, m. As indicated by Eq. A-12, the coefficients of the exponential curve can be generated by taking the log of the dependent variable, Y_i, before insertion into the regression equation. Again, a_2 will be the antilog of the intercept, a, and $b_2 = m$.

A warning should be noted that, in the process of linearlization, scales have been changed and thus, deviations unequally weighted. Thus the least squares regression method is no longer strictly correct.